DISASTER AND RESCUE ON CANADA'S
ATLANTIC SEABOARD

A Sea King helicopter crew prepares to fly from the deck of HMCS *Halifax* to search for survivors from the MV *Leader L,* which sank March 23, 2000, about 700 kilometres northeast of Bermuda. Thirteen people were eventually saved; eighteen died. (DND)

Deadly Frontiers

Disaster and Rescue on Canada's Atlantic Seaboard

DEAN BEEBY

GOOSE LANE

Edited by Laurel Boone.
Cover illustration © CP Photo Archive (Joe Gibbons).
Book design by Ryan Astle.
Printed in Canada by AGMV Marquis.
10 9 8 7 6 5 4 3 2 1

Canadian Cataloguing in Publication Data
Beeby, Dean, 1954-
Deadly frontiers: disaster and rescue on Canada's Atlantic seaboard

Includes bibliographical references and index.
ISBN 0-86492-311-2

1. Search and rescue operations — Atlantic Coast (Canada)
2. Canada. Canadian Armed Forces — Search and rescue operations — Atlantic Coast (Canada)
3. Disasters — Atlantic Coast (Canada) I. Title.

HV555.C3B44 2001 363.34'81'09715 C2001-901736-7

Published with the financial support of the Canada Council for the Arts, the Government of Canada through the Book Publishing Industry Development Program, and the New Brunswick Culture and Sports Secretariat.

Goose Lane Editions
469 King Street
Fredericton, New Brunswick
CANADA E3B 1E5

In memory of Matthew,
who slipped from our grasp

Contents

Introduction

A ship sinks, a plane crashes, a child wanders deep into the forest. Death is imminent, its grip almost inevitable except for the bravery and persistence of small groups of men and women who enter these dark frontiers as rescuers. They fail sometimes. But often they return with the near-dead, plucking them from the hungry jaws of disaster. The aura of death gradually dissipates in the cabin of a rescue helicopter, in the back of a bouncing ambulance or on the deck of a heaving coast guard ship. Everyone — rescuer and rescued — has dreams for the rest of their lives about these taut moments. Electric dreams, about peering across the dreaded frontier.

Tales of disaster and rescue are deeply embedded in the modern consciousness, often elevated to myth. The saga of Sir John Franklin's 1845-1847 expedition, two forlorn ships lost somewhere in the Arctic, has fascinated generations not only because of its unresolved mystery but because of the misadventures of the many expeditions that have searched in vain for Franklin's crews. Robert Falcon Scott's mad dash to the South Pole in 1911-1912; RMS *Titanic*'s killing descent in 1912 to the bottom of the North Atlantic; miners trapped in a collapsed Nova Scotia colliery in 1958 — all of these conjured life-and-death accounts that continue to fascinate in ways far out of proportion to the original loss of life. Modern civilization seems to crave stories about death's numbing frontier, partly as a reminder of the transitory nature of this life and partly for the inspiration that survival can provide.

Most of our planet has been explored, some of it over-explored, yet humans still feel compelled to test their personal limits on ocean floors, on mountaintops or across vast ice fields. But there's another deeply human impulse: to tour the boundary between life and death, a demarcation line that can appear without warning in the heart of a city or in the midst of an ice barren. The geography is incidental. It's a drifting, ephemeral zone that crackles with fear. Rescue workers know this special place well, return to it often, and do so for the noblest of reasons: to snatch those caught at the disintegrating frontier and return them safely to their families.

Canada occupies a unique position in the rarefied world of search and rescue. The second-largest country on the planet, Canada has three jagged coastlines, an immense internal wilderness and a vast Arctic to swallow hapless travelers. At the same time, Canada is among the most thinly populated regions on earth. With minimal state support, Canada's search-and-rescue communities have been forced to make the most of inadequate resources and equipment across broad seas, forests and tundra. The result — miraculously — has been a tightly focused, efficient and effective culture of search and rescue that has much to teach the world. Indeed, that expertise has become a little-known cultural export. And in an era of growing recreational adventure and increased trade and travel, with its attendant risks for disaster, Canada's lessons are needed now more than ever around the globe.

Since the Second World War, and more intensely in the last twenty years, Canada's East Coast has been the crucible for modern search-and-rescue techniques and equipment. This hard-won regional experience has been driven mostly by disaster, from the 1982 sinking of the *Ocean Ranger* oil rig off Newfoundland to the disappearance of numerous cargo vessels in the 1990s, including the *Protektor*, the *Gold Bond Con-veyor*, the *Marika 7* and the *Vanessa*. Ground search and rescue, a special branch of this culture, was reborn in 1986 during the protracted search for a lost boy in the forests north of Halifax. In 1998, Swissair Flight 111 plunged into waters off St. Margaret's Bay, Nova Scotia, triggering a massive search-and-recovery effort as well as a fundamental rethinking of emergency response. And one of the worst disasters of the last twenty years within the search-and-rescue community itself was the 1998 crash of a Labrador helicopter from Canadian Forces Base Greenwood, Nova Scotia, leaving six rescue specialists dead among the charred wreckage. Crashes, sinkings and lost-persons cases do happen elsewhere in Canada and have challenged crack rescue teams in other regions. But they occur more frequently and severely along the Atlantic seaboard, where major air and sea routes intersect with powerful weather systems, and where the deep bush beckons at the edges of villages, towns and cities. Canada's east-coast triumphs and tragedies, in turn, have helped forge a professional search-and-rescue culture that is second to none.

Media reports on Canadian search-and-rescue missions are often strong on drama but woefully weak on depth and accuracy. Reporters understandably seek out the "hero" in a rescue story at the expense of the skilled teams that work in the shadows. All rescues draw on the hard work of hundreds of men and women who are rarely mentioned in newspapers or newscasts, from the scientists who study human endurance times in cold environments to the coast guard specialists who draft maps estimating the drift of victims in sea currents. Their vital contributions remain largely uncelebrated, even though many are regarded as world leaders within the rescue professions. The long list of Canadian search-and-rescue achievements includes specialized search-plotting software, satellite hardware and advanced rescue beacons and buoys, as well as lost-persons theory and sophisticated techniques of search management. The "heroes" or front-line professionals certainly have their place in the pages that follow, but they are joined here by the people who design, support and run the system behind the scenes. They, too, enact compelling stories of tenacity, dedication and even bravery.

Each account in *Deadly Frontiers* is centred on a few major east coast search-and-rescue missions, moving from the land to the sea and air. Ground search and rescue, as will be seen, grew rapidly in the space of just a few years from an amateur, often bumbling exercise by well-meaning volunteers to an effective and efficient service run by trained semi-professionals. The transformation came about only through the heartbreaking death of a young boy in the backwoods of Nova Scotia and the frustrations of the 6,000 searchers who tried to find him. This landmark case — now cited in search-and-rescue seminars across North America — establishes the larger theme for the rest of the book. Significant advances in marine and air rescues have come about only through the catalyst of major tragedies that expose weaknesses in planning, training and equipment. The sinking of the bulk carrier *Flare* and the crash of Swissair Flight 111, both in the watershed year of 1998, each demonstrated weak links in planning and equipment, forcing a wrenching re-examination of assumptions. Later that same year, the fiery crash of an aging Labrador helicopter raised the most troubling questions of all: had the penny-pinching federal government been playing with the lives of its search-and-rescue crews by delaying the purchase of replacement helicopters?

The crash of Labrador 305 illuminates the second major theme of *Deadly Frontiers*. Canada's elite rescue corps come by their expertise in part because many are lifelong outdoor enthusiasts who have learned from personal experience to deal decisively with the extremes of weather and environment in this northern maritime nation. But their extraordinary performances in difficult rescues also result partly from the chronic unsuitability of the equipment and resources with which they are provided. These men and women bridge the gaps with a wide range of outdated technology, from geriatric helicopters to archaic radio systems. They endure long, jarring trips to rescue scenes because their bases are so scattered and few. They risk physical exhaustion trying to overcome the limitations of their second-rate personal equipment. The Labrador 305 disaster is a grim warning that, despite their ingenuity in making the best use of inadequate technology, Canada's rescue specialists are themselves sometimes caught at the deadly frontier. That tragic lesson is still not well understood outside the world of search and rescue. If this book has any value, it will be to help remove politics and false economies from decision-making in government and allow Canada's much-admired search-and-rescue culture to evolve rationally, without the persistent need for disasters and tragedies to open purse strings.

The men and women profiled in these chapters were generous in allowing me to enter their world and courageous in speaking honestly about their vocations. We met and spoke in hangars, cockpits, cargo holds, passenger cabins, wheelhouses, amphitheatres, club rooms and offices. They allowed me to observe field training, classroom instruction, briefings, debriefings and missions-in-progress. All ranks, from recruits to veterans, talked openly about their hopes and fears, frustrations and satisfactions. There were no conditions attached to my access, and none of the people I interviewed asked to review the manuscript. These are their unvarnished stories. Their missions failed as often as they succeeded, but never for lack of tenacity. Many of these professionals return again and again to the bleak frontiers out of duty and passion, with a perspective on the world that few of us can ever imagine.

CHAPTER ONE

Groundpounders

Andy Warburton and the Rebirth of Ground Search and Rescue

Ground search and rescue in Canada was reborn on a Tuesday afternoon in the summer of 1986 at the edge of Rasley Meadow, a boggy area just north of Square Lake, Nova Scotia. The spot is less than three kilometres west of the tiny rural community of Beaverbank, itself a half-hour drive from Halifax. The houses at Beaverbank, like those across much of Nova Scotia, are within a stone's throw of pristine wilderness. Walk a few hundred paces and you're enveloped by thick, tangled bush. There are no landmarks, no visible trails, and you'd better have a compass — a sense of direction is the first thing to disappear in the woods. Nova Scotia has been called the lost person capital of North America because all over the province, dense forests lie waiting to swallow victims just beyond their secure homes. Dozens of people get caught each year, even on the outskirts of Halifax, the biggest urban centre in Atlantic Canada. It had happened yet again, in tiny Beaverbank, in the summer of 1986.

A group of expert Mi'kmaq trackers had been through Rasley Meadow on the Saturday. They'd been called in from Kentville, in the Annapolis Valley, to help with the search, but then they were asked to go home again. Not needed, they were told by organizers at the battered old bus that was being used as a command post. It was the second time that week the Mi'kmaq had been called for no reason and the second time they'd been peppered with tasteless insults from some of the people milling about. But this time the native trackers refused to leave, saying they would conduct their own search. You'll be arrested, they were warned. We'll take that chance, they answered, and set out across the Beaverbank River toward the northwest. Experts estimate that a human being leaves at least 2,000 potential clues to his passing for every kilometre travelled, not counting the millions of particles of shed skin that a trailing dog can trace. Some hours later, the Mi'kmaq group radioed back that they had hit the jackpot of clues — they had found the prints of small running shoes. The organizers at the command post, angered at the outright defiance, sternly ordered them out of the area. This time the frustrated Mi'kmaq complied.

The next visitors to Rasley Meadow were Canadian army reservists from Camp Aldershot, also in the Annapolis Valley, who arrived three days after the Mi'kmaq team. The group was part of a 400-strong military contingent that was conducting massive sweep patterns across previously searched areas. By mid-afternoon Tuesday, a few of them stumbled on dirty sneakers whose sole patterns seemed to match the prints the Mi'kmaq had found. The grey Squirt-brand shoes had gotten wet, likely in a nearby brook that fed Square Lake, and had apparently been carefully laid out to dry. A couple of hours later, two militia members found the owner of the sneakers, a slight, blond-haired nine-year-old boy wearing a blue-and-white T-shirt and green shorts over his bathing suit. Curled up in a narrow deer run, he looked peacefully asleep, but it was the sleep of the dead. He had been lying motionless like this for at least three days, and he had been wandering, frightened, for up to five days before that. He died, cold and alone, just over two kilometres from his worried mother, father and twelve-year-old brother, from a warm bed, a bright kitchen, hot meals and big drinks. Somehow this little boy had eluded the largest ground search ever mounted in North America.

Word of Andy Warburton's heartbreaking death from hypothermia flashed across Canada within fifteen minutes of the discovery of the frail, barefoot body. The search, which had been launched late on Canada Day, Tuesday, July 1, had grown into a major media event. Millions of Canadians worried about the lost child in Nova Scotia. How far could a small boy have wandered? So many searchers — up to 1,200 at the peak on Saturday and more than 5,800 throughout the operation. How could they all come up empty-handed? Just who was in charge, anyway? And why wasn't the military called at the outset? These and other vexing questions began to consume the parents, the relatives, the reporters, the police, the politicians and the searchers themselves even before Andy was buried in a tearful ceremony in his hometown of Hamilton, Ontario. For months afterward, agonized organizers retraced their steps, gnawed by the tragedy. Many of the searchers began to complain publicly about the botch-ups. The parents demanded an official inquiry. A television special mercilessly documented the gaffes. Key search leaders began to resign their posts, unable to continue their volunteer service in the face of blame for this disaster. Their best intentions, their sleepless nights and aching backs, their hunger and headaches, all of it had blown up in their faces. Canadians were calling them incompetent.

In those dark days during the summer of 1986, few could imagine that a small miracle was in the making. Few could foresee that Andy's sad end was about to trigger a revolution. One boy's lonely death would rumble out from Rasley Meadow like a ground tremor, reaching across a continent.

■ ■ ■

The story of Andy Warburton began as a typical summer vacation in Nova Scotia, far from the smog, humidity and congestion of southern Ontario. The Warburton family had been visiting Helen and Jim Bulger, Andy's aunt and uncle, for two weeks in quiet Beaverbank, nestled beside a clean lake about twenty kilometres north of Halifax. Andy's father Tom described his son as a happy-go-lucky kid who hated school and wanted to be a goalie. Helen, Tom's sister, found him quiet and a little nervous. Being an urban kid from Hamilton, Andy was not "woods wise" by any stretch. But he liked swimming, and as his mom Doreen and aunt Helen cut firewood in the yard of the Bulger house, he asked whether he could go for a swim at the Tucker Lake beach, a short walk away. Since his brother Gary was already at the lake, cooling down on a warm summer day, Doreen, of course, said yes. Andy changed in the house and was gone. It was the last time Doreen would see her son alive.

About an hour later Gary returned without his brother, saying Andy had never arrived — words that jabbed like an icicle into his parents' hearts. The family and neighbours immediately checked all the likely places: the house, the back yard, the neighbour's yard, a short road north that ended at a creek where Andy had once been covered with leeches. Nothing. A little over an hour later, they rang the local detachment of the Royal Canadian Mounted Police, and a sympathetic woman officer was soon at the house jotting down details. She eventually found a neighbour's child, twelve years old, who claimed to have seen Andy crossing that leech-infested creek into some dense woods as he called out Gary's name. This spot, just to the north of the Bulger house, was then declared the "point last seen" or PLS, the reference point that would serve as the basis for all subsequent search planning. That PLS would be marked on dozens of maps, surrounded by concentric circles, by grids and by pins. Straight-edge rulers and pencil tips would be lined up on this point, and index fingers would here begin their slide through topographic lines of elevation and across the blue lines of rivers. This reference point would even guide the decision about where to establish the command post. Police dogs would begin their tracking at this spot. From here, mantrackers would set out looking for broken twigs, crushed leaves and bent grass blades.

Because the point last seen orients the activities of everyone involved in a search, identifying it correctly can speed up a search tremendously. Unfortunately, the Warburton point last seen was almost certainly wrong — the first of many dubious decisions that turned the operation at Beaverbank into a prototypical example of how not to conduct a major ground search. The RCMP officer, for example, noted that the neighbouring kid probably saw Andy at about two o'clock, but Andy had asked permission to go to Tucker Lake later that afternoon, perhaps around three-thirty. The creek was also north of the house, whereas Tucker Lake was located to the southeast, along a short trail. On the face of it, taking the kid's information as identification of the PLS just didn't make sense. The officer nevertheless accepted the report as the only sighting, without seeking further for either corroborating or conflicting evidence. Many weeks after the tragedy, a neighbour told a reporter she had actually seen Andy heading east to the lake at about four o'clock. This neighbour had never been interviewed by the policewoman or by any other search officials, and she had assumed, incorrectly, that those in charge already knew what she knew. And so, before a single searcher had entered the woods, the operation had gotten off track.

As is standard procedure, the RCMP called in a local volunteer search group to begin scouring the heavy forest north of the creek. The Waverley Ground Search and Rescue Team had been formed in 1972 in the wake of another disastrous search for two children. On a Wednesday in late February 1969, a nine-year-old girl and her six-year-old brother went missing in the woods next to their house at the southwest outskirts of Halifax, in a new subdivision known as Leiblin Park. Typical of Nova Scotia, the dense bush lay like a trap just beyond the well-tended lawns of suburbia. Halifax police called on the neighbourhood for help, and soon parents, teenagers, kids and a local Rover troop were massed in the thick bush, tromping alongside burly city cops. Many of these volunteers were poorly dressed for the weather, as thick snow began to shroud the ground, and none had been given anything but vague instructions to watch for the lost pair. "Look for the untrodden areas," Ollie Robinson, Halifax's chief of police, told the searchers. "Look under every branch, look under every fallen tree. Look around every boulder, and inside every cave." Schoolgirls in nylons and teenage boys in thin jackets wandered unsupervised.

At two a.m., searchers did find the boy, Murray Graves, wet and crying on top of a boulder a bare 500 metres from his home. But his sister Rhonda was nowhere to be seen, though one volunteer did claim to hear her calling out to her brother. The next day, a Thursday, the call for help widened. Halifax's deputy police chief announced, "We want everybody we can get — everybody." Soon about 200 navy personnel joined concerned citizens and neighbours to scour the woods, the forest floor already honeycombed with the tracks of searchers. Among the new arrivals was Stephen Oakley, a twenty-two-year-old man who was at least familiar with the bush. Despite the increased manpower, there was no sign of Rhonda that day, either. And the next morning, Halifax police had two missing-person cases on their hands: Oakley, an avid hunter and woodsman, had disappeared as well.

The search Friday was mainly by "conga line," a primitive and awkward technique in which a line of people spaced perhaps two arm's-lengths apart sweeps across an area. Groups of fifty people combed differing areas at first. These volunteers were then combined into a giant chain of 300 people, each spaced more than a metre apart. After an hour of fumbling, this deep-woods chorus line produced no results, not even a clue. Three helicopters brought in for the search also found no sign of either the girl or Oakley. A separate group of three volunteers finally located Rhonda's body at about a quarter to four, at the end of a line of small snowprints. The little girl had died of exposure, lying under a tree a mere kilometre from her home. "I saw a few indentations in the snow that looked like footprints, so I followed them," said David Casey, the man who located Rhonda's body. "When I saw her, she looked like she was asleep under the tree."

Oakley's body was found the next day. Poorly dressed in thin pants, a thin shirt, a sweater, boots and parka, he, too, had died of exposure. "Stephen was lying on his back with one hand up to his face, as if he were blowing on his fingers to keep warm," said one of the men who found him. "His glove was between his knees, and his head was covered by a parka." Oakley had entered the woods with a friend from the south to test a personal theory about Rhonda's whereabouts, but then he chose to carry on alone, quickly becoming disoriented and hypothermic. With more than 2,000 poorly organized volunteers involved in the search, it was a miracle there were no other deaths or injuries.

The Graves-Oakley disaster triggered some soul-searching about how, without some basic organization and training for the searchers, a simple lost-person case can quickly escalate into a larger tragedy. "When the old system was used, they'd arrive at a scene, and somebody would put a map on the hood of a car and say, 'You guys go this way, you guys go that way,'" says Tony Rodgers, who in 1969 was one of the eighteen-year-old Rovers who joined the search. "It showed us that we needed something more than a police officer leaning on the hood of a car."

The province's Emergency Measures Organization, prodded by the Graves-Oakley disaster, soon offered workers' compensation and small annual grants to encourage the formation of volunteer search-and-rescue teams. The organization also produced a short "manual" with advice on conga lines, on emergency equipment and the use of search dogs. The pamphlet also urged a check-off system to ensure that every volunteer sent into the woods was accounted for at the end of the day. "This helps eliminate the possibility of someone being left in the area, which could result in having another lost person," it advised. The manual's search tips today sound naive and amateurish, but they at least represented a first step in what was to become a long and difficult learning process. Among the earliest in Nova Scotia to take up the challenge were a handful of members of the Waverley volunteer fire department, who decided to branch off into this relatively new business of search and rescue, becoming only the second such group in the province. The priority was training. Oakley's unexpected death had highlighted the dangers of sending hundreds of unschooled volunteers into the woods. It was a hard lesson that would need repeating seventeen years later in the Warburton search.

■ ■ ■

The Waverley group, with its command-and-communications bus, arrived at the Bulger house at seven-thirty at night to begin the search for Andy Warburton, several hours after he disappeared. (Most organized ground searches begin at night, since missing persons are usually reported to police after they fail to turn up by sunset.) The search director immediately went to the RCMP officer on the scene for instructions; all searches for missing persons are considered criminal investigations with ultimate responsibility residing with the police agency of jurisdiction. In Nova Scotia, as with many of Canada's less-populous provinces, the RCMP polices the smaller communities and rural areas. The point last seen was

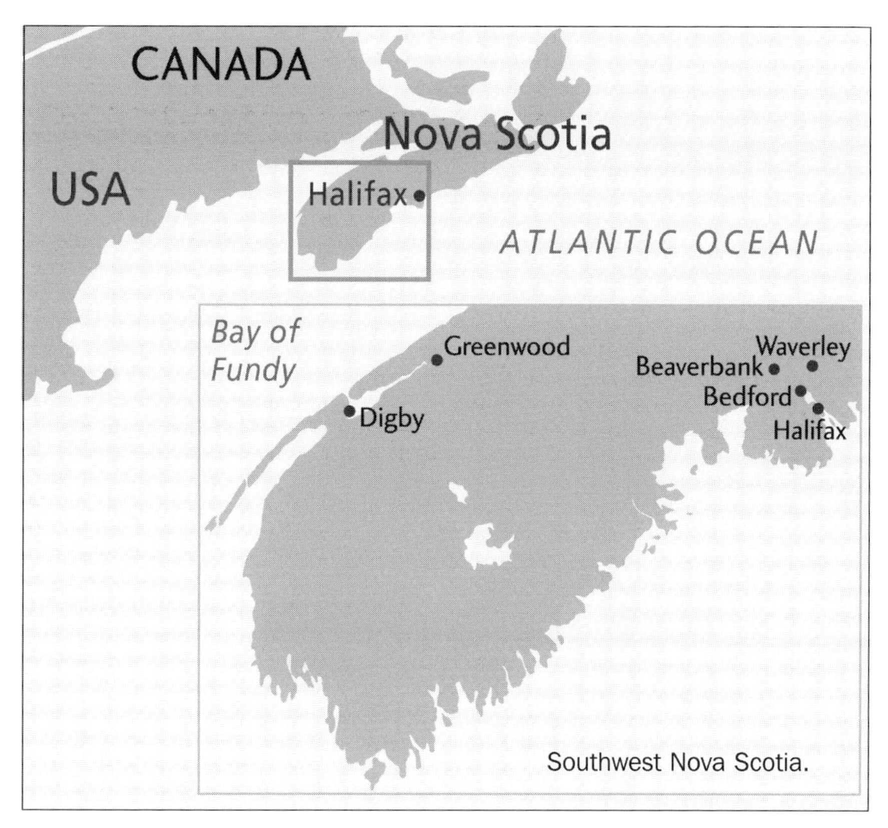

Southwest Nova Scotia.

identified, and soon forty-four searchers who had given up part of their Canada Day holiday entered the woods north of the creek, divided into four teams. Despite the darkness, they remained in the bush until well after midnight as the temperature dipped to an unseasonably low eight degrees Celsius. The teams conducted grid searches, basically the standard conga line in which members follow imaginary lines through the woods regardless of the obstacles, each person separated from the next by a distance of several metres. The search director met the worried parents that night to reassure them, but he did not solicit any more information about Andy's disappearance from neighbours or other potential witnesses.

The operation was suspended at three-thirty and resumed at dawn, about two hours later. At eight o'clock, a forestry helicopter provided by the provincial government began to sweep the search area as a trickle of unsolicited volunteers arrived at the command bus. By suppertime, the Waverley organizers were swamped with several hundred well-meaning people offering to help — so many, in fact, that several dozen were turned away. But television and radio reports kept drawing in waves of citizens, from teenagers to seniors, many forced to mill about the bus as those inside coped as best they could with an unwanted, untrained army. Meanwhile, organized teams from other areas of the province arrived after calls for help. RCMP divers poked through the reedy silt of Tucker Lake. That Wednesday afternoon, a light drizzle turned into cold rain, which would last for eight solid hours. Still no sign of Andy, missing now for more than twenty-four hours.

Desperation seems to have set in on Thursday, when the Waverley officials shifted the entire search area seven kilometres to the east and headquartered it in a local school. The basis for this decision was a flurry of dubious sightings: newspaper delivery boys who thought they saw Andy run back into the woods after they called his name; a woman whose camper trailer door had been opened and a child's toy deposited inside; an RCMP officer who thought he'd caught a glimpse of a boy who looked like Andy. Seven kilometres was an awfully long way for an ill-dressed little boy to travel through tangled bush, especially considering the unseasonably cold nights, the pounding rain and the more than 400 searchers watching for him. But the idea took root that a frightened Andy was deliberately avoiding his searchers, perhaps being driven farther and farther away from the Bulger house. In some respects, it was a plausible theory: children are taught from a young age to avoid strangers, especially men who bid them come near. Some children can also mistakenly feel they've done something wrong by losing their way and may irrationally try to avoid discovery to escape possible punishment. To forestall this possibility, Andy's mother was brought in to tell him over a megaphone that she wanted him home, that he should trust the searchers. "Mommy and Daddy are waiting for you. We love you, Andy," she called. "The men will not hurt you. They are here to help you. Please come out on the road."

The search organization began to crumble under the strain. Although the search director and others had been working at the scene from the very start on Tuesday night, the Waverley team was not about to turn over leadership to another group, even on a shift basis. That would be an ego-bruising admission of weakness and failure. Experience has since demonstrated that clear-headed thinking in such situations begins to wane after eight hours and disappears after about twelve hours, leaving organizers muddled and ineffective. The body's signals are unmistakable: complexions flush, there is stammering and memory loss, and the person in charge often has a far-away look in his eyes, routinely failing to respond to simple questions as if words can't penetrate the mental fog.

Compounding the fatigue in the Warburton search was weak co-ordination among the search director, his helpers, the RCMP officer nominally responsible for the overall operation and the head of the province's Emergency Measures Organization, who became involved as the scale of the search escalated beyond anyone's previous experience. Inevitably, contradictory reports began to be issued, and sometimes-vital data was not disseminated properly. People watching TV at home often had more precise information about Andy's physical appearance than many of the trained search teams at the site. Worse, the mistakes of the Graves-Oakley disaster were being repeated. Untrained, spontaneous volunteers were being sent into the woods without adequate controls. When some of these people did not check in after their assignments, trained searchers had to be sent into the woods to find them, wasting precious resources. Fortunately, no one became truly lost — but the incident was one more signal that the search for Andy was stumbling badly.

The search was called off at eight p.m. on Thursday as the cold rain pelted the forest canopy and meadows. Instead of ordering another night search, the RCMP sent about a dozen men into the woods with microphone dishes that would amplify the sounds of a child walking or whimpering. But the experiment failed despite hours of patient listening, and Andy's ordeal in the autumn-like conditions crossed the forty-eight-hour mark. His parents, meanwhile, pleaded with the head of the Emergency Measures Organization to send in the military. There was no shortage of manpower, officials assured Tom and Doreen, and their impassioned request was set aside. A clutch of reporters at the site soon became a conduit for the rising anger and frustration of the searchers, as teams complained bitterly about the incompetent organization. The grumbling got so bad that there were fears of a riot in the school gym serving as a temporary barracks, and RCMP officers were called in to calm things down.

The next morning, Friday, things got even more desperate. Organizers conjured up a search technique they had never used before, dubbed a "staggered blanket search." The approach was like a conga line, except that each human link preceded the next one into the woods in a kind of bizarre, Rockettes-style cascade. The advantage was never clear or explained, and confusion reigned. The approach was soon abandoned in favour of the standard grid search that had been used without success since Wednesday. Don Bower, then a radio expert attached to the Waverley group, well remembers being part of a thirty-five-person conga line covering old territory in the driving rain. The downpour was so loud that directions could not be shouted to everyone but had to be passed on by word of mouth down the line. After about an hour, everyone was soaked to the skin, and the search was halted. "I knew things weren't going right, and I didn't know why," Bower recalls. "Everyone appeared to be doing everything they possibly could, but it just didn't seem right."

In the leadership vacuum created by this chaos, psychics inevitably made their appearance and were given a major role in the search — another sign of the growing despair. Modern search teams are trained in a broad range of hard sciences, from mathematics and advanced navigation to radio technology and emergency medicine. They generally sneer at psychics, seers and clairvoyants, often with the put-down that "if you know so much, why didn't you warn everyone that this person would get lost?" But these practitioners of parapsychology are not necessarily dismissed outright. The family of the lost person may be comforted by the assurance that all avenues are being investigated, so the psychic may be tolerated for humanitarian reasons. A search director may have some unwanted and untrained volunteers on his hands in need of a make-work project. Having them chase down a psychic's predictions can solve a very real personnel problem. Witnesses to the circumstances of lost-person cases may want to avoid personally embarrassing questions by presenting themselves as psychics. An astute search director will be able to spot this ruse and accept the genuine information with no questions asked.

The organizers of the Warburton search had no such excuses when they allowed a handful of psychics to direct operations temporarily. These overwrought officials, plainly strung out after more than three days of ineffective activity, were ready to listen to even the wildest of predictions. At least two psychics were flown over the search area by helicopter to try to get a "fix" on Andy's location, suspending ground operations for about three hours. A few teams were asked to check the validity of some of these extrasensory perceptions about Andy's whereabouts, while others were joined by the psychics themselves. The most bizarre incident occurred when a psychic from elsewhere in the country was able to persuade a search co-ordinator by telephone to go to a location and await further telepathic instructions. The co-ordinator complied but for some reason was unable to receive the ethereal instructions and abandoned the exercise. Needless to say, none of these forays into the paranormal brought any of the searchers closer to Andy. "There were hours of lost time due to the psychics," members of the Colchester search team complained afterward. "We see no reason for stopping all the men and women for hours while these psychics play their games."

By Saturday, with no sightings of Andy, primary search operations were switched back to the area of the Bulger house. More people were off work that day, and the number of volunteers swelled to about 1,200. An RCMP helicopter was brought in from Fredericton, and more divers checked waterways and lakes. The Mi'kmaq trackers arrived and found the sneaker tracks — the first genuine clue in five days of searching — but they were dismissed and the clue ignored. The organizers, at wits' end, finally bowed to the request of the relatives and called in the troops. Defence Minister Perrin Beatty telephoned the parents at about eleven that night and offered men and equipment, whatever was needed. The catch was that it would take until Monday morning for the first contingent of 400 soldiers to arrive, half of them regulars and the other half reservists. The discipline and organization of the military was doubtless a balm for the relatives and the frustrated searchers, who had watched the command structure wither inside the battered Waverley bus. Never mind that the infantry had no training in basic search techniques and was relatively slow to respond. Things could get no worse, and the military at least had a hierarchy and command structure that offered the closest thing to hope.

The rain and fog finally receded on Sunday, and temperatures rose above twenty degrees Celsius again. If Andy had made it through the bitter cold and could warm himself up on Sunday, the soldiers might yet save the day. The search area was now confined to a circle about one and a half kilometres around the Bulger home, and many of the 800 searchers conducted their own renegade sweeps of the woods, now thoroughly fed up with organizers. Like the Mi'kmaq, they ignored threats of arrest. Areas that had initially been considered unlikely search zones, such as woods on the opposite side of highways or settlements, were now examined. A couple of heavily silted man-made ponds were drained. Still no Andy. The military finally showed up on Monday morning and reported to the Waverley bus. "On arrival at the Search HQ, for the initial orders, it quickly became apparent that EMO (Emergency Measures Organization) had never conducted a search of this magnitude before," Major R.R. Burns wrote later about his first encounter with the haggard, red-eyed civilians who were nominally in charge of the operation. "It was expected that Search HQ would be better organized — with a smooth running HQ Search Cell. This was a disappointment and is a major criticism . . . all operations at Search HQ appeared disorganized, i.e., maps would be on the wall one day and on the floor the next."

The military units proceeded to use a sweep search pattern that covered all of the previously searched areas around the Bulger house, then expanded the coverage by several kilometres. Andy continued to elude them through Monday and into Tuesday morning, and few now held out hope he would be found alive after seven straight days and nights in the bush. Searchers tacitly acknowledged that this had become a search for a body. The discovery in Rasley Meadow on Tuesday afternoon by two militia members confirmed everyone's suspicions. The terse radio log entry for 5:25 p.m. read: "Found subject in C1 condition," or Charlie One, the Waverley team's special code for dead. The only piece of code the team ever used, it would keep bad news away from radio eavesdroppers until family members could be notified. It's a code the Waverley group hoped never to need.

Andy's behaviour during his final hours will never be known for certain. But subsequent to the search, more has been learned about a lost child's mental state. Children of this age begin to experiment with so-called short cuts, some of which can actually be longer than direct and familiar routes. These short cuts often get them in trouble, and Andy may have been following what he thought was a novel route to Tucker Lake. When children in this age group are disoriented, they tend to become panicky and irrational, especially if they're alone. Many will run some distance along trails, far from the point last seen. This behaviour may be the metabolic by-product of fear, which increases the supply of adrenalin and pumps more blood into the legs, preparing the body for flight. Few such children in this predicament will successfully find their own path back to civilization — virtually all of them have to be found. In addition, it's not uncommon for kids of Andy's age to hide from searchers out of fear. Toward the end, as the latter stages of hypothermia set in, Andy may have removed his sneakers because his feet felt too hot. Known as paradoxical disrobing, this frequently observed phenomenon is common among lost people of all ages. Hunters, for example, have been known to strip down and jump into frigid lakes as their bodies generate a last burst of heat before succumbing to the cold. Shoes and pants are usually the first to be shed. Andy's sneakers, which had been described as laid out to dry, may in fact have been removed to help his small, tired feet cool down.

Recriminations over the disaster began even before the autopsy results were available. The fact that the military was successful when the Waverley crew was not lent credence to the family's insistence that soldiers would instil discipline and direction into the operation. Why were they not recruited at the outset? Some teams questioned why Rasley Meadow had been deemed low priority throughout the seven-day search. The response was that helicopters and a few searchers had indeed checked the area of Rasley Meadow, but Andy was likely hiding in the early days and then was partially under the brush when he finally succumbed. There were urgent calls for an official inquiry from Hamilton city council, from the family, from the province's chief medical examiner and from some of the Nova Scotia teams themselves. "The conduct of the search . . . displayed a level of incompetence that bordered on negligence," wrote a member of the King's County search-and-rescue team in calling for a judicial review.

The Nova Scotia Cabinet steadfastly resisted the pressure. Instead, officials convened a July 30 meeting with all the search participants that produced a rather lame document outlining the need for better co-ordination in large, multi-team searches. The statement did allude to the problems of having a search director work long hours without a break. "A clear mind is needed at the search director's desk, and one individual working to exhaustion will not help find the missing person." But the seven-page declaration was otherwise full of platitudes. "This document can only be viewed as a not very subtle attempt to sweep an unpleasant and important issue under the carpet," a searcher from another team wrote later to the head of the Emergency Measures Organization.

The criticisms grew increasingly public with the broadcast on November 27 of *Eight Days in July*, a half-hour CBC documentary on the search-gone-sour. This well-researched piece vividly recounted the disorganization in the field, and a reporter found the Bulger neighbour who for the first time revealed that she had seen Andy heading east that fateful afternoon, not north toward the creek. The CBC program drew the plausible conclusion that Andy had accidentally veered left instead of right at a fork in the path leading to Tucker Lake, eventually getting caught in the bush to the northwest. The CBC researchers also suggested that supposed sightings on Day Two of the search may in fact have been deer, and that reported sounds of a child might actually have been the noises made by breeding porcupines, which can readily be mistaken for children's voices. Adding fuel to the debate was a lengthy article in the April edition of *Cities*, a short-lived Halifax-Dartmouth magazine, that was highly critical of the disorganization and missteps of the Warburton operation. The diary-like account, by journalists Kelly Shiers and Stephen Kimber, quoted at length many of the disgruntled members of other Nova Scotia teams who were called in to help. The grumbling always seemed to focus on the Waverley crew rather than on the RCMP, even though the Mounties had made the questionable snap decision about the point last seen, and even though they had final authority.

The Waverley team itself issued a confessional statement a few weeks after the search, saying its planned worst-case scenario — a ninety-six-hour search with up to 1,000 volunteers — was simply overwhelmed by the Warburton operation, which took 165.5 hours and involved more than 5,000 volunteers. The group went on to list eight areas where it fell short. "We made inadequate use of non-S[earch] & R[escue] volunteers, including incomplete briefings, screen, control in and out of the woods, and inefficient use of available search effort," said the statement. "We were inadequately prepared to provide logistical control, monitoring, documentation, documentation retrieval and quality control of the organized assisting S&R teams." But the self-flagellation did little to forestall the criticism. The second-guessing continued, and even today there are older members of other teams who hold grudges against the Waverley crowd for the disaster of 1986. If heads were demanded, they were eventually delivered voluntarily. Almost every person in authority who co-ordinated the Warburton search has since left the emergency field, including Bruce Golding, director of the Waverley team. "I think all of us, probably for all the wrong reasons, still feel guilty to this day," says Don Bower, a Waverley radio operator. "I can't drive by Tucker Lake without just for that fleeting second wondering if the outcome could have been different." Instead of seeking professional help for their personal anguish, "some of the fellows debriefed at the tavern," says Bower. One man connected with the search committed suicide several years later, and many of those who knew him say Andy's death weighed heavily on his mind.

■ ■ ■

At some point during that black July weekend, someone from the RCMP placed a telephone call to Ken Hill, then a forty-one-year-old child psychologist at Saint Mary's University in Halifax. Hill, a quiet, genial man, was born and raised in the United States but came to Canada in 1971 as a post-graduate student to study psychology at the University of Alberta in Edmonton. A Saint Mary's professor since 1976,

Ken Hill. CH

Hill toiled in traditional areas of child psychology. He had never worked alongside the police and had no special knowledge of search and rescue, though he did drive an ambulance for the U.S. Air Force between 1966 and 1970 and was required to have some paramedic skills. Hill had read about the Warburton search in the newspapers and had seen it on television, but he never imagined he might have a role to play. The Mounties were now asking him to come to the Beaverbank search headquarters to help them understand exactly what a nine-year-old boy does when lost. Hill dutifully arrived — but didn't have a shred of advice to offer. "I felt so embarrassed about it," he recalls. "I knew of no research on it." All he could draw upon were his own memories of being a scared nine-year-old boy lost in the Santa Monica Mountains, north of the Los Angeles suburb of Venice, where his family then lived. After half a day of wandering, he climbed onto an outcropping and spotted his father driving slowly along a road below. He waved, was seen and saved. But now, as a child psychologist, Hill knew nothing of any practical use to help find another nine-year-old boy who was likely just as frightened as he had been thirty-two years earlier.

Hill well remembers the chaos at the elementary school that became an assembly point for the Warburton search. "It was insane," he says. "There were hundreds of people milling about in small groups, some visibly angry or upset. . . . There was no indication that anybody was in charge." Being of no use to the police, Hill offered himself as a volunteer searcher; he was initially rejected until a Saint Mary's student spotted his psychology professor and invited him to accompany his team. The conga line was the only search method known to and used by the Waverley group. "You just went to the point last seen and did an ever-expanding circle around that until you finally caught up with the victim," Hill recalls. Experience has shown that no supervisor in the hierarchy of a search should have to manage more than seven people, yet field leaders in the Warburton search were lining up dozens of volunteers for grid searches that — not surprisingly — produced no results.

The discovery of Andy's body that Tuesday at Rasley Meadow was emotionally wrenching for everyone, not least for Hill, whose years of training in child psychology had been worthless in the search. A week later, Hill drove out to the Waverley team's training facility at Lakeview to fill out a membership application. And in the weeks that followed, the professor became an attentive student of Ron Marlow, the group's chief training officer and himself a biology professor at Dalhousie University. Hill also read voraciously. One of the first volumes he absorbed was Tim Setnicka's *Wilderness Search and Rescue* (1980), which drew on earlier research by an obscure Canadian, Bill Syrotuck, who had died in 1976. Marlow eventually showed his new protegé a battered, tenth-generation photocopy of a Syrotuck work, a thirty-five-page pamphlet called *Analysis of Lost Person Behavior: An Aid to Search Planning*, published by Syrotuck's widow in 1977. Hill flipped through the pages with mounting excitement, as if he had stumbled across the Dead Sea scrolls. Syrotuck's analysis had the very answers that Hill was unable to give the RCMP that July. The slim volume would radically alter the course of his career.

William George Syrotuck was born on March 13, 1930, in Edmonton, Alberta, where he was raised as an only child in a Ukrainian household. Soon after graduating from high school, he joined the Canadian army and became an intelligence officer with the artillery corps. He fought in the Korean War, then returned to Edmonton. In late 1956, soon after leaving the military, he married Jean, a nursing student from Cadomin, Alberta. With Jean, he moved to Calgary, where he took an electronics course, then to Montreal and a job with Canadair in 1958, and finally to Seattle, Washington, where he worked for Boeing on a two-year contract to develop the Bomarc missile. In 1961, Syrotuck joined the University of Washington as an electronics technician, first in oceanography and later in applied physics.

Syrotuck's interest in the world of search and rescue was sparked in the early 1960s by a news report of a lost child about fifty kilometres north of Seattle. The family dog had alerted rescuers to the location of the child, and the incident prompted Syrotuck to explore the possibilities of using dogs to find lost people. He and his wife taught their own German shepherd how to locate missing people using air-scent techniques, and they joined like-minded dog owners at a local club, participating in more than 200 dog-assisted rescues in the Seattle area over the next decade. In 1972, Syrotuck left the university to do some writing, producing his first search-and-rescue book, *Scent and the Scenting Dog*, which today is considered an early classic in the field.

Seattle during this period became a hotbed for new search-and-rescue techniques, being an area of high population density adjacent to mountainous wilderness where people frequently became lost. As the Syrotucks developed expertise with search dogs and became part of the growing search-and-rescue community, they were drawn to the larger problems of search planning. Mathematically inclined, Bill Syrotuck decided to apply statistical analysis to records he obtained of 229 people lost in wilderness areas in the United States. Most of the incidents had occurred in his own Washington State (117) and in New York (95), with a handful of incidents from seven other states. His premise was that a wandering person can quickly create giant search zones that cannot be covered in a reasonable time by even a large number of searchers. But if search organizers can better predict the behaviour of the wanderers, then the search zone can be confined to a more manageable area. "Predicting a likely location is valuable for several reasons," Syrotuck wrote. "First, there are generally not enough resources for a search of all the possible places or areas. Second, although subjects of the search may eventually be found, they may no longer be alive. It is, therefore, paramount to find them quickly."

In his study, Syrotuck divided his 229 cases into eight categories, including hunters, hikers and the elderly. Of special interest to Ken Hill was the category of children aged six to twelve years. In a chart based on the records of searches for children in this age group, Syrotuck calculated that a third were found within a mile (1.6 kilometres) of the point last seen, and three-quarters within two miles (3.2 kilometres). Hikers and hunters, on the other hand, tended to travel farther. The charts made no reference to the actual distance traveled by the lost person over twisting terrain, since there was often no way to be certain. Instead, the calculations were all as the crow flies, and they were directly applicable to the Warburton search: Andy's body was found inside the same high-probability zone that Syrotuck had identified almost a decade earlier. Members of the Waverley team had possessed this vital information for years but never properly applied it to their searches.

Syrotuck also dabbled in psychology, though it was not his forte. He observed that the six-to-twelve age group, "when detached from their familiar surroundings, such as outings with the family or other groups, or from adult-sponsored activities . . . can quickly become lost. They have been transplanted into strange environs by their parents or other adults. They blithely follow along in the activities without realizing that there may be a need to keep track of their own whereabouts." Children in this group also readily sought shelter in bad weather, often squirming next to or inside logs or snuggling against the sides of boulders. Syrotuck went on to note that more than two-thirds of the lost children between these ages had followed paths, game trails, streams or drainages rather than plunge into thick bush. "When lost, they are likely to be beset by all the fears and problems of adults, but with a greater sense of helplessness or loneliness, because of their lack of experience," he wrote. Syrotuck's study, published shortly after he died from heart surgery, went on to suggest detailed search strategies based on such factors as whether the subject was on flat, wooded terrain or in a hilly area. "I was just amazed," Ken Hill says. "I was a believer right then."

Here was an empirical, practical approach to quickly developing search plans customized to the lost person. One needn't have a detailed psychological profile of the missing individual, only the basic demographics and some indication of the activity, whether hiking, hunting or wandering. As Hill put it later, "Land search began the difficult transition from an art to a science with the publication of Syrotuck's pioneering monograph." *Analysis of Lost Person Behavior* was, of course, only a mathematician's attempt at psychology based on limited information. A trained psychologist could surely take the analysis much further, to increase understanding of the mental mechanisms at work when humans become disoriented and must survive somehow. At the very least, Hill realized, data on lost persons in Nova Scotia could be collected and analyzed in the same way to augment and extend Syrotuck's work, providing a localized template for conditions prevailing in the province, three-quarters of which is forest. Syrotuck's approach could be applied to the very next ground searches in Nova Scotia. These were the challenges Hill set himself as he read through this seminal study.

But Hill soon discovered that Syrotuck's contribution to search and rescue was broader still. Earlier in the 1970s, Syrotuck had begun to apply an exotic branch of mathematics known as probability theory to the problem of ground searches for lost persons. The intellectual genealogy of this approach stretches back to the Second World War, when U.S. mathematician Bernard Koopman developed statistical approaches to tracking down enemy vessels such as submarines. The methods were carried over into peacetime by the U.S. Coast Guard to help locate lost ships at sea. But only a few individuals, Syrotuck foremost among them, had understood how useful this analytical approach could be in trying to locate a lost person in the wilderness. Two other Syrotuck volumes, *An Introduction to Land Search Probabilities and Calculations* (1975) and *Some Grid Search Techniques for Locating Lost Individuals in Wilderness Areas* (1974), laid out sophisticated ways to cover territory quickly. Rough and ready conga lines, then standard procedure in many ground searches, were exposed as inefficient and unreliable. "The chances of detection remain high only as long as the system remains completely organized and under control," Syrotuck warned. "As disorganization, or lack of control, develops, the chance of detection rapidly becomes lower." The words, almost a template for the Warburton disaster, had never been properly understood or heeded in Nova Scotia.

Hill immediately began to collect lost-person statistics for Nova Scotia, contacting the Emergency Measures Organization for permission to inspect their files. He found that some useful information could be extracted from incidents as far back as 1976, though record-keeping was generally inconsistent and sloppy. Hill eventually devised a new form that was distributed in 1988 to the province's twenty-three search-and-rescue teams; it has become the basis for an annual crop of statistics he harvests each spring. From these numbers, Hill constructed a chart based on more than 250 lost-person incidents, divided into nine categories: hunters, fishermen, hikers, miscellaneous adults, youths aged thirteen to fifteen years, children aged seven to twelve years, young children ages one through six, despondents and walkaways. The chart lists the range of distances from the point last seen that members of each group were located, and it calculates the zone where half of them were found. This high-probability zone is intended as a practical guide to help search managers decide where to throw their resources at the start of an operation. Andy Warburton's body was found just where the chart predicts, inside the fifty per cent zone for his age group, as Syrotuck's chart had also predicted. Indeed, the striking thing about Hill's localized chart is how closely it resembled Syrotuck's charts, which drew on data from nine states. The similarities suggested something universal about lost-person behaviour, even in differing wilderness areas.

The Waverley team members, meanwhile, carried out several more searches in late 1986 that again badly exposed their disorganization and ineffectiveness. Mary Ellen Parsons, four, and Jonathan Dominique, five, became lost in the bush near their home on September 9, 1986, at Lawrencetown, not far from Halifax. About fifty team members — Ken Hill and Don Bower among them — were called to the scene at about eight p.m. and spent the next eight hours laboriously grid-searching while neighbours and family also scoured the area. The pair were found safe at four a.m., asleep in each other's arms, in a gravel pit just half a kilometre from their home. Mary Ellen's dog was with her and barked loudly, alerting searchers. Non-team members made the actual find. The operation should have concluded hours earlier, and the discovery should not have depended on a dog's bark. On November 3, the team was called out to search for an overdue deer hunter after his worried wife alerted police. Jacques Champoux, twenty-six, became lost in the bush near Powder Mill Lake, north of Halifax, wearing light clothing in temperatures that dipped below freezing at night. Police located his car on the east side of a road, and the ground search, involving more than 200 people, was concentrated in the territory to the east. After two days, searchers began to examine the west side and heard a gunshot near the lake late at night. The next morning, the team located discarded articles of Champoux's clothing — boots, jacket, vest, cap — a clear sign of paradoxical disrobing, the by-product of severe hypothermia. Five months later, Champoux's naked body was found floating in the lake. The search team surmised that he had been drinking alcohol in the woods and slipped unaware into hypothermia because his inebriated body did not shiver, vital feedback that he might be in trouble. He apparently stumbled for a day or two in an irrational state and entered the lake to cool down as his numb body gave a last burst of heat. "That was another travesty," says Hill, who joined the search. "The guy died within throwing distance of the team's base."

The Champoux death was the last straw for a few frustrated members of the Waverley team. "We had a whole bunch of bad searches before we said, 'This is enough,'" says Hill. Bernie Marshall, search director for the Warburton disaster, was re-elected to the post for 1987, an ominous sign that it would be business as usual for the Waverley team, which numbered over 100 people. But a group of about five team members, worried there would be no fundamental changes, launched a quiet revolution. They formed a so-called search management committee to learn about new techniques they had all read and heard about from other places in North America. Hill and the others, for example, began to study manuals developed by the National Association for Search and Rescue, based in Chantilly, Virginia, and to apply them to new searches in Nova Scotia — often without the involvement of Marshall. The group also arranged for an expert from the association to come to the province to teach the team some of the proven approaches to search management, and they persuaded the RCMP to pick up some of the travel costs.

Among the first changes for the Waverley group was adoption of the "incident command system," an approach to search management originally developed for U.S. forestry services to help fight forest fires. Some of the system's premises were that the chain of command should be clear, transparent and co-ordinated; that key responsibilities should be delegated; that communications should be integrated and terminology made common; that all activities should be properly documented; and that work should be carried out within limited shifts. Searches are managed by a small "overhead team" or "general staff," different members taking care of logistics, operations and plans and all of them answering to the incident commander. This system was a far cry from the "search boss" approach the Waverley team had used for years, with one person in charge until the end and with uncertain delegation and unclear responsibilities down the hierarchy. In the Warburton search, it might more properly have been called the "search bosses" approach, with as many as five managers giving sometimes conflicting directions. Haphazard management can work tolerably well for small teams for years, but inevitably it will break down in chaos. "The longer it works, the harder it is for you to accept a new way of doing it," says Hill.

The Waverley rebels also began to experiment with more effective alternatives to conga lines. Grid searching still has a role in modern searches, especially when the victim is thought to be unconscious or unlikely to answer calls — a child, for instance, or someone with Alzheimer's disease. But the method is rarely a first approach. Long lines of searchers can be highly inefficient, especially when individuals are closer than about fifteen metres to one another. These human strings move relatively slowly at a time when speed is essential, and they can destroy vital clues, especially if the line is made up of untrained, spontaneous volunteers. The preferred first approach is to call in highly trained human trackers and hasty teams. Trackers quickly chase down clues where a lost person has disturbed the route, whether by churning the soil or bending a twig. Hasty teams are small, mobile groups of searchers also trained to spot clues and determine the lost person's direction of travel. Both trackers and hasty teams will use a sophisticated bush technique called "binary sign cutting," which can quickly establish a route and help limit the search area by determining the places where a lost person has not been. Typically, these experienced searchers will leapfrog over one another as they quickly identify a trail without necessarily checking every clue along the route. Heavy rain, winds or snowfall can limit their effectiveness, but trackers and hasty teams are now considered the shock troops of any modern search operation. In the year after the Warburton disaster, they began to appear in the woods of Nova Scotia.

Ironically, as Hill began to develop fresh techniques for ground searches, he himself became lost with his family in an urban park. Hemlock Ravine, just off a highway north of Halifax proper, is a heavily wooded oasis veined with brooks and meadows. At its entrance is a heart-shaped pond, built to commemorate a nineteenth-century royal romance, where large numbers of yellow-spotted salamanders mate early each spring. In behind is a series of pretty trails that meander aimlessly, some of them cutting back into each other, others winding deep into the forest where they peter out. The trail system has been left incomplete for years, and a poorly drawn map of the area posted at the entrance is constantly vandalized, often missing altogether. Hill took his young family — his wife Carol and their two boys, aged four and eleven — for an outing there early in 1987 on the recommendation of a friend. The park lulled everyone into a false sense of security, with suburban houses visible at points and the terrain appearing serene and secure. The Hill clan wound up on a circular trail and discovered they had somehow looped around it a second time without spotting any clear means to exit. Hill found what appeared to be a route out, but the trail proved to be a dead end deep in the bush. The family then had to bushwhack east through rough terrain along the ravine itself. They finally emerged onto the busy Bedford Highway, the road they had taken to get to the park. "It was in winter, so it was a non-trivial experience," Hill recalls in a typically understated manner. He became lost yet again on another visit with the two boys. This time they followed the ravine west until coming upon another highway. Hill was temporarily disoriented a third time when visiting the park alone and was about to bushwhack out when he recognized a familiar trail that led him out.

Hemlock Ravine, which Hill describes as an environmental trap, thus became his field laboratory. "It's a fantastic place to study lost person behaviour," he says. He visits two or three times a month throughout the year, sometimes with a group of students and others who become his subjects. "I began losing my friends and SAR (search and rescue) colleagues there, following them around as they tackled this forest labyrinth, taking copious notes as they confronted each bewildering intersection," he says. More often, Hill goes to Hemlock Ravine alone. "On nice sunny days, I can often be found sitting on a park bench near the beginning of one trail, knowing that through the course of the afternoon several people will be asking me for directions," he says. "They're spatially confused. I'm very friendly to them because I know that when they start to head down one particular trail, they're lost." Hill offers some face-saving comments, especially to men, who generally dislike asking for directions. Everybody gets lost here, he assures them, even me. Most of the lost visitors exhibit a quite typical behaviour: they refuse to reverse direction on a trail they've been walking on for a while. Instead, they press ahead through the tangled undergrowth, often emerging onto one of the two highways. Hill frequently accompanies those who do ask him for directions out of the park as he talks to them about their experience of spatial confusion.

Beginning in 1991, Hill also developed computer-based 3D mazes at Saint Mary's for testing college sophomores to help him understand how the human brain navigates unfamiliar territory. As a member of the Waverley ground search-and-rescue group, the busiest of the province's teams, he also had access to a steady stream of found persons. Many of these consented to an interview, explaining what got them lost in the first place and how they had behaved in the wilderness. The interviews, the statistics, the field lab and the computer mazes all helped Hill to develop a much clearer picture of lost person behaviour, to expand upon and correct the errors in Syrotuck's earlier analysis. His cutting-edge research has provided a highly practical guide to search teams who must quickly develop a plan of operations.

One of the first of Syrotuck's categories that Hill revised was the "elderly persons" group, those above sixty-five years of age. "Grandpa (who may be a little senile) has been left at the car and has disappeared by the time everyone else appears," Syrotuck wrote in 1977, referring to a family outing. "Or Grandpa joins in the 'picking' and has a small fall (unnoticed) that leaves him shaken and confused. When someone checks for Grandpa, he has disappeared." In twenty-four cases that Syrotuck studied involving lost persons over age sixty-five, most (fifty-seven per cent) were found within a mile (1.6 kilometres) of the place last seen. In fact, Syrotuck had incautiously lumped in people suffering from Alzheimer's disease and other dementias with all other elderly people. As Hill demonstrated in a 1992 paper, lost hunters, hikers and fishermen over the age of sixty-five are capable of travelling just as far as their younger counterparts. His study of 338 cases of lost Nova Scotia hunters found that the lucid elderly can negotiate their environments just as well as the young. Indeed, the lost older person appears readier to build a shelter and await rescue than the young, increasing his chance of survival. Syrotuck's error was assuming that dementia and age went hand in hand, and it badly skewed his calculations of how far this group typically traveled. "Such 'ageism' was perhaps excusable in 1977, before old age and senile dementia were widely revealed not to be synonymous, as research indicates that only a fraction of non-institutionalized elderly appear to have a dementia-related cognitive impairment," Hill wrote in one of his research reports. Lost dementia victims, he argued, should be part of a larger category called "walkaways," regardless of age.

Walkaways are today a growing problem in the search-and-rescue world. The label refers to anyone under constant care in a home or hospital for severe mental problems, whether psychosis, retardation or senile dementia. (The group excludes so-called despondents, people who are clinically depressed or suicidal and who take to the woods for solitude or to kill themselves.) As public health-care services are cut and more such patients are de-institutionalized or given greater freedom of movement, more of them are winding up in the woods. Walkaways have become the fastest-growing lost-persons group in ground search and rescue, supplanting hunters, who have been disappearing far less frequently thanks to wilderness education programs. Of some 300,000 Canadians with Alzheimer's disease, about seventy per cent are classified as chronic wanderers. Hill's statistics demonstrated that walkaways do not travel far from the place last seen but are nevertheless hard to detect because they usually don't answer the calls of searchers or cry out themselves. Some even hide. Many do not even realize they are lost.

Recent research has suggested that many wandering Alzheimer's victims suffer from what has been called "motion blindness" — an effect similar to walking around with their eyes closed — rather than from memory problems. Some Alzheimer's disease patients have been known to walk in a straight line into dense underbrush or even into a lake, where they quickly drown. They rarely stick to pathways or trails. They are rarely dressed for the weather, many clad only in pyjamas. Their behaviour often mimics that of lost small children except that they have a high mortality rate in the bush, forty-five per cent in Hill's analysis of twenty-two such cases in Nova Scotia. Another study in Ontario by James Hanna examined twenty-two cases between 1981 and 1996 in which patients of health-care facilities were found dead after having gone missing. A third were eventually found on the grounds or in the facility itself. One man was found in the building twelve days after having gone missing, dead of dehydration, even though the premises had been searched by staff. Unlike other groups, the behaviour of walkaways is highly unpredictable, presenting a vexing problem for ground search teams. Hill's research suggested that search dogs should be brought in early and that thorough grid searching may be necessary at the outset because the victims often hide in deep underbrush. Their high mortality rate calls for urgent action.

Another of Syrotuck's categories that Hill clarified and expanded upon was the "despondent," the depressed or suicidal individual, also highly problematic for search organizers. Syrotuck found only five such individuals in his nine-state sample, too few to create a chart. But he did note that all "went upward to some distinctive location: an open area with a view, beside a picturesque lake, or the top of a scenic hill. They tended not to go into underbrush." Hill found sixteen cases in Nova Scotia, many of whom were found where different types of terrain meet, such as a secluded meadow next to a thick forest. (Much of the province is relatively flat, which may account for the lack of corroboration for Syrotuck's observation about high, scenic locations.) Few travelled very far, and they rarely answered the calls of searchers. Many had drugs or alcohol in their system. Not surprisingly, the mortality rate was quite high — sixty-five per cent, either from suicide or exposure. This spoke to the urgency of such cases, though search-and-rescue members are understandably reluctant to become involved when the lost person is known to be suicidal and has a weapon.

Small children, ages one through six, often become lost while following an animal into the woods, whether a dog, a deer or a raccoon, Hill determined. The youngest in this age group do not even have a concept of being lost. "They 'don't know that they don't know' their way around the woods, and it apparently doesn't seem to occur to them that they could get lost," Hill has written. "This probably accounts for why young children (quite unlike the school-aged child) are rarely afraid of becoming lost per se . . . and will readily follow an animal into the forest or strike out on little exploring expeditions, paying no concern to the return trip." They tend to seek shelter in bad weather and will sleep at night rather than wander. They do not travel far, but none has ever been known to respond to searchers' calls. Nevertheless, they generally have an excellent survival rate because of their sheltering and sleeping instincts. Older children, those about eight years old, begin to create mental maps as they discover that the world has a constant structure independent of their travels. "One realization that now occurs to them is that the route they use to travel between locations is arbitrary: there are, indeed, many paths to Rome," Hill found. "Frequently this insight provides the seed for a new curiosity about their environments, particularly a fascination with the very concept of a short cut." These so-called short cuts can be longer and more dangerous than the familiar routes and can get them into trouble. Andy Warburton, as we have seen, may have been a victim of such curiosity.

Lost hunters are perhaps the best understood group; this has historically been the most populous lost-persons category. Ego can seriously complicate a search, since most hunters refuse to admit to being lost, only "turned around." Or, as Hill has put it, for hunters "lost" is a four-letter word. Many will travel great distances to try to extricate themselves from the woods without help, and a third of them will succeed.

Hill's research went on to analyze the idiosyncratic behaviour of lost hikers, fishermen, youths and others. One vexing category of lost person is the "bastard search," as it's known colloquially among Nova Scotia teams. These are incidents in which the missing person is not lost but is pretending to be or is deliberately avoiding friends and family because of emotional problems. "I have found no other search incident so likely to end up in a false search as that involving teenage girls who have entered the woods in pairs, particularly if one of them has recently had some sort of emotional encounter with a parent or a boyfriend," Hill has written.

Hill tried to plug a gaping hole in Syrotuck's study, that is, whether or not there was a difference in behaviour between lost males and females. The Nova Scotia data indicated that there was no significant difference in the straight-line distance travelled, though relatively more women survived. This was especially pronounced among the despondent category, where proportionally more males than females were found dead — paralleling suicide statistics in general that show males succeed in killing themselves far more often than females. The Nova Scotia numbers also showed that far more men than women became lost, though the ratio is skewed by the fact that virtually all hunters were male. About a quarter of lost children were female, as were more than half of all despondents.

The Nova Scotia statistics helped illuminate group dynamics in the woods, a factor that Syrotuck had ignored even though between a third and a half of all incidents involve more than one lost person. Hill showed that lost groups of adults, usually consisting of two or three people, travelled the same distances as individuals and were as likely to survive as those lost alone. But he also found that lost children under five years of age frequently separate from their companions, while those over five — including teenagers — tend to stick together. These older children behave more maturely and rationally when they are lost with a companion than when alone. Interviews Hill conducted with survivors also strongly suggested that people of all ages were far less frightened and behaved more rationally when in groups.

Hill's mountain of lost-person research was remarkable not only for its breadth but also for its direct applicability in the field. Other social scientists have since expounded on individual categories such as lost Alzheimer's patients as these grow rapidly in number. Ed Cornell and Don Heth, behavioural scientists at the University of Alberta, have studied lost children in urban and suburban settings. Indeed, their work was sparked by a Warburton-like incident when a nine-year-old boy went missing in July 1979 in a wilderness park in Western Canada. The RCMP called asking for help, but there was nothing to offer. They were shamed by the Mountie's reply: "Don't worry, Doc, we may get a psychic out here today." The child was never found, and the experience prompted some basic research about lost child behaviour. Even so, Hill stands virtually alone in this area of inquiry. His detailed distance-travelled chart for lost people in the Nova Scotia setting has replaced Syrotuck's version, rather than localizing it. No other such chart exists for any other place in North America, and ground search teams around the world today draw their probability circles based on the experience of a small corner Canada's Atlantic seaboard.

Nova Scotia has also become a model jurisdiction for the organization of ground search and rescue, thanks in part to the Waverley team's rebirth. Hill was elected search director for 1988, after his intense seventeen-month apprenticeship, and stayed on until 1996 to ensure the new techniques and approaches became permanent. (He remains an active member of the team to this day.) Many of the other teams that formed in the wake of the 1969 Graves-Oakley search, however, continued to use differing techniques and terminology even after the Warburton search. "Every team had developed its own standards, its own direction, its own search methods, its own maps," says Hill. "They had been allowed to develop independently." The Warburton disaster clearly demonstrated the dangers of multi-team searches in which the participating units used widely diverse operating procedures. Therefore most teams across the province joined in 1989 to create the Nova Scotia Ground Search and Rescue Association, a body intended to provide common training standards that would allow a search director from one team to slip seamlessly into the command of an operation when another team's director had completed a twelve-hour shift. Instruction in the Incident Command System was begun for the 1,550 members of all teams, and forms, procedures and radio routines were all eventually standardized. Every square centimetre of the province was made the responsibility of one of the twenty-three main teams. And an RCMP officer was assigned to spend forty per cent of his time working to promote and organize ground search and rescue. Unlike in the United States, where many public officials — including sheriffs — are elected, appointed police officials in Nova Scotia provide a needed continuity and consistency. Indeed, the relationship with the police is now so close that search groups are regularly called upon to help find criminal evidence, such as discarded weapons, in the woods.

The decade-long transformation of search and rescue in the province was not without its pain. Many of the pre-Warburton search groups were partly social in nature, modeled after hunting clubs, without a commitment to professionalism. Many of the Waverley rebels were not welcomed by the old guard as they tried to shed the traditional approaches to search and rescue. "People were talking on the one hand about mission statements, and other groups were saying, 'Are we going to have a dance Saturday night or not?,'" says Don Bower, who was one of those pressing for change. "A lot of us took a lot of abuse as we tried to build a better system." Fellow rebel Mike Murray, who had been a spontaneous volunteer for the Warburton search before signing onto the Waverley team, remembers the defensiveness of long-time team members. "It was an old boys' club," he says. "I can remember being threatened by people who would phone up and say, 'We're going to get you if you don't back off.' We were rocking the boat." As Hill recalls, "There was a lot of fighting and bickering between the new guard and the old guard." The fault line appeared at times to run along socio-economic lines, with blue-collar members clinging to the past while the new group, made up largely of white-collar professionals, insisted on radical change. Provincial bureaucrats were also slow to respond to the rebels' demands and were sometimes obstructive. Bower, Hill and others realized early on that only constant pressure from below would get anything accomplished. Two of the Waverley rebels, Hill and Ron Marlow, were easy to dismiss within the Nova Scotia search-and-rescue community because both were university professors with American roots. Bower, though, a Waverley team veteran, had respect in the province. It was his personal support that helped give the rebels the credibility they needed.

The Waverley group, recently rechristened Halifax Regional Search and Rescue, has more than 130 members today and is the busiest in the province, responsible for about twelve of the sixty to seventy lost-person cases each year. Training is constant and intense — a favourite proverb at their Lakeview facility is, "The more we train, the luckier we get." The group's motto is also a reminder that a new discipline has replaced the unco-ordinated amateurism of the past: "We serve as one — that others may live." New members are placed on a six-month probation to ensure they make the grade. Only about $1,000 of the team's $30,000 annual budget comes from taxpayers, with the remainder drawn from time-consuming fundraisers. They have some beat-up vehicles, one of them a bus with a collapsible radio tower, as well as an ancient medical truck and an all-terrain vehicle. The bus serves as a mobile command, control and communications centre, patched together with surplus military equipment, volunteer labour and about $10,000 in donations. The rag-tag group, which has traditionally drawn hunters and fishermen, now includes computer aficionados, radio experts and ordnance officers. Some key members go to searches toting briefcases filled with forms rather than compasses. Women make up about a quarter of the current membership and the numbers are climbing. The province estimates the volunteer services of all these groups are worth at least $2 million a year, but the teams easily doubled that figure when they spent 48,000 hours searching the shores of St. Margaret's Bay for human remains from the Swissair Flight 111 crash in September of 1998. Members have learned to live with media reports that frequently credit the military for work they have quietly performed.

Nova Scotia has begun to export its newfound expertise. Shortly after the Warburton search, the province's Emergency Measures Organization promoted a program to "woods-proof" children, and it quickly spread to Andy's hometown in Hamilton. An estimated 30,000 children and their parents have been through the course. The team is also accustomed to visitors from other provinces and countries wanting to find out how to set up successful units within their own jurisdictions. Ironically, the half-hour CBC television show that so stung the Waverley members in 1986 has now become a basic training tool. Copies of *Eight Days in July* have been shown in search-and-rescue classrooms across North America to demonstrate the dangers of amateurism. Like zealous missionaries, the Waverley rebels themselves pledged to take their newfound expertise throughout the continent to prevent other tragedies, and Don Bower and Mike Murray have been especially active in this regard. But Hill himself has been the main catalyst for taking the lessons of Rasley Meadow to the rest of North America. He has edited a collection of seminal essays on lost person behaviour, sponsored by the federal government, and he dedicated the volume to Andy Warburton. In 1992, he became a director of the National Association for Search and Rescue, the American body that acts as a training centre and information clearing house.

Hill, who has dual Canadian-American citizenship, also teaches for the association, training the trainers as well as students, and in 1997, he was invited to substantially rewrite and edit the body's main training manual, *Managing the Lost Person Incident*, used all over North America and in many other parts of the world. This is an update of the very manual the Waverley rebels had absorbed in the fall of 1986 to revolutionize the culture of search and rescue in the province. The new version opens with a chapter on the Andy Warburton case, which has literally become a textbook example of how not to conduct a major ground search. A section on lost-person behaviour condenses a decade of Hill's research in Hemlock Ravine, in the Saint Mary's computer lab and in the forests of Nova Scotia. His distance-travelled chart is reproduced, eclipsing Syrotuck's groundbreaking work. The chart's statistics for lost children includes the distance that Andy managed to travel — just over two kilometres — before he died. "Too often, the realization that good management is critical to the success of the search incident comes only after a prolonged and disorganized search which has ended tragically for the lost person," Hill reminds the reader in the introduction. Writing the manual, he says, was a labour of love, a way to honour the still-fresh memory of a nine-year-old boy.

Andy Warburton's parents today feel satisfied with the outcome of the controversy they helped stir up over their son's death. "We know some good came of it," says Doreen, who moved to Stoney Creek, Ontario, with her husband Tom shortly after the tragedy. "We're sure of it." The Warburtons have since visited Beaverbank, Nova Scotia, again but admit to some unresolved bitterness about the search for their son. "We appreciate what the people of Nova Scotia did but, unfortunately, we didn't get what we should have got in the first place," says Doreen. "As much as their hearts were there, it was wrong." But, of course, the team today would likely find a little lost boy like Andy in less than a day.

The Waverley group's hard-won expertise has been adopted throughout the province and has spread to neighbouring New Brunswick and across the country to British Columbia, where there are a significant number of wilderness searches. But the lessons have been slow to arrive elsewhere in Canada. Sadly, some provinces have had to repeat the bitter experience of Nova Scotia. In Saskatchewan, for example, a futile two-week search for an eight-year-old girl appeared to duplicate many of the errors in the Warburton case. Ashley Krestianson became lost in unfamiliar country near Tisdale on July 14, 1994, when she separated from her twin sister to take a short cut — typical behaviour for her age. Her decomposed body was found by a hunter many weeks later. A pathologist ruled out foul play, suggesting hypothermia as the cause of death. During the frantic first days of the search, the police were overwhelmed with more than 300 volunteers whom they could not use effectively. Unsubstantiated sightings of Ashley prompted organizers to move the search area farther away from where her body was eventually discovered, about seven kilometres from the point last seen. After three days, search leaders realized they had been looking for the wrong shoeprint. Soldiers, none of whom were trained in ground searches, were brought in after five days to augment the volunteers. A thirteen-metre experimental blimp, developed at the University of Waterloo, was brought to the scene by a Sault Ste. Marie, Ontario, search-and-rescue team to play Ashley's favourite music on loudspeakers from the air and perhaps draw her out. The helium-filled, remote-control device was the brainchild of a Toronto lawyer, Norman Loveland, who had watched the tragic ending of the Andy Warburton search and sought a solution in technology. The frustrated RCMP eventually called in a hypnotist who, working with the twin, hoped to tap into some kind of mental telepathy with Ashley and thus discover her whereabouts. About a hundred untrained volunteers formed into a massive conga line, more than half a kilometre long, controlled by one team leader hollering into a bullhorn. Tragically, Ken Hill and others from the Waverley group offered their help in the first days of the search but were quietly rebuffed by the Saskatchewan RCMP.

■ ■ ■

Brandon Gray, seven, and his older brother T.J. Jewett, twelve, may not know it, but they likely owe their lives to Andy Warburton and the lessons his death taught to a generation of ground searchers. Like Andy, Brandon and T.J. had come from out of town with their mother to visit the home of an uncle. The uncle and his wife had recently moved from Ontario to a house at Whites Lake, Nova Scotia, a rural community just outside of Halifax where — typically — the houses back onto the bush. On the afternoon of Saturday, March 20, 1999, just as winter was becoming spring, the Cape Breton brothers wandered out of the back yard and into the forest, where they quickly became lost. "We were just walking a little bit and turned back to come back and couldn't find nothing," T.J. said later. "It was all woods." It was about two o'clock, and, like Andy, the pair were poorly dressed for the environment — dark, thinly insulated clothing and sneakers. Neither had been woods-proofed, and they kept travelling rather than remain stationary and "hug a tree," as a popular kids' safety program teaches. The family and neighbours initially tried to find the boys, but at five p.m., as evening approached, they called the RCMP's Tantallon detachment to report Brandon and T.J. missing.

A municipal police officer who lived in the area, meanwhile, began a search using his air-scent dogs, the most common type of dog used in search and rescue. These are trained to locate the human scent and follow it back to its source. Much rarer are trailing dogs, trained to follow only the scent of a particular person after having been given a sniff of clothing or other personal belongings, such as a pillow case. The scents both kinds of dogs follow are created by the natural shedding of fifty million human body cells each second, most of them skin cells. These discarded cells, known as "rafts," combine with other bodily excretions such as sweat and oil to create personal scents as unique as fingerprints, scattered on the forest floor or caught in an eddying air current. The dogs in the Whites Lake search eventually followed two airborne scents, but they turned out to belong to untrained volunteers who were in the bush looking for the boys. This common problem can be overcome by ensuring that the search area is kept searcher-free for half an hour or more. But it turned out that there was almost no wind that evening to carry any human scent to the noses of the dogs. Neighbours and dogs would be of little use — the boys would need a team of trained ground-pounders, as the search volunteers are known today. So a call went out shortly after seven p.m. to Tony Rodgers, search director of Halifax Regional Search and Rescue, the modern incarnation of the old Waverley group.

A long-time search-and-rescue follower, Rodgers had volunteered as a temporary searcher during the Graves-Oakley and Warburton disasters but had since officially signed up with the Waverley group. The pager call from Whites Lake came, ironically, while Rodgers was at a nearby outdoor recreation show, manning a search-and-rescue booth to promote the Halifax group and solicit volunteers. He was at the Whites Lake home about half an hour later, about the same time of day that the Waverley group began the Warburton search almost thirteen years earlier. He eventually called in thirty-one members of the team to the Whites Lake address, and, as darkness descended, there was an intense door-to-door canvassing of the neighbourhood to determine whether anyone had seen the children. Family members were questioned in detail, then questioned again. The uncle's home was finally declared point last seen, and five so-called hasty teams, with three members each, were sent to check out segments of the immediate area to see whether they could establish the boys' direction of travel. Unfortunately, the woods were still contaminated with renegade searchers, and the hasty teams had a few false alarms as neighbours continued to wander and call out. Rodgers reluctantly sent an all-terrain vehicle to quickly scout out a bush road that the boys might have followed. ATVs have the advantage of speed but can inadvertently destroy vital clues — and modern searches are all about locating clues as much as people; as some have put it, lost people themselves become the ultimate clue. With no sign of the two kids, Rodgers asked for reinforcement teams from five other districts. A provincial helicopter was put on alert for morning, since the choppers cannot fly at night. And an RCMP forward-looking infrared unit, or FLIR — which can spot the wisps of heat given off by the human body — was ordered brought in from Fredericton, the nearest detachment that had one.

At the battered command buses, which were set up at the local Legion, the team cartographer placed an overlay on top of a detailed 1-10,000 map. A standard distance-travelled chart compiled by Ken Hill was used to determine just how far a lost child might travel. For children between the ages of seven and twelve, experience in Nova Scotia had shown it could be as much as eight kilometres or as little as just under one kilometre. However, half of all lost children had been found in the zone between 1.6 and 2.73 kilometres from the place last seen. Accordingly, two circles were drawn to these distances around the uncle's house, giving a rough outline of the primary search zone. Into this zone, segments were partitioned based on the local topography. A river or "blue line" might set the limit on one segment, a highway on another. Experienced search teams were then assigned to segments and were given detailed descriptions of the boys, their clothing and footwear. They were even handed an image of the sole on one of T.J.'s Adidas sneakers. An alert police officer had noticed that a magazine on the floor of the mother's car had been stamped with a muddy heel-print, which was redrawn, measured and distributed. Sure enough, one of the searchers positively identified a track shortly before midnight, a distinctive diamond within a diamond. The track, to the west of the house, was next to a set of smaller tracks, Brandon's.

Spontaneous volunteers flocked to the bus offering to help. It being a Saturday night, the first question put to each of them was: "Have you been drinking tonight?" Anyone answering yes, or anyone even smelling of liquor, was turned away. Others who were inadequately dressed, without flasks of water, flashlights or compasses, were also rejected. And no one was allowed into the woods except under the supervision of trained searchers. At the height of the search, about 150 people meticulously checked an area of several square kilometres. Finally, at 3:24 a.m., a husband and wife team from the East Hants search-and-rescue group found the boys at the edge of Nine Mile River. Like Andy, T.J. had taken off his wet sneakers and was apparently considering crossing the swollen river when he was finally discovered. The spot was just inside the outermost circle drawn on the search map. "We were lying down, and I had his feet in my jacket," T.J. said after the ordeal. "I took my shirt off and wrapped it around my feet because my feet were wet. We kept by each other [for] body heat." The boys had been wandering for about thirteen hours but were in remarkably good shape. The paramedics said all they needed was a hot bath. "You're a couple of tough Cape Bretoners," Rodgers told them when they were brought back to the base. Brandon assured everyone that they hadn't been frightened. The relieved, tearful mother praised her oldest son's perseverance. "I'm so proud of T.J. He kept moving," she said. "He thought if he stopped that nobody would find him." In fact, T.J.'s mobility probably prolonged his stay in the woods. Even so, the boys had been located safely within eight hours of Rodgers's arrival at the home, well inside the first twelve-hour shift the teams set for themselves. Ken Hill himself was booked in for the next round, beginning at eight a.m. But these days, the group does not often require a second shift.

The next night, about forty members of the Halifax team gathered in the old white schoolhouse at Lakeview, near Waverley, for the standard debriefing. The main room has a checkerboard linoleum floor. The drab walls are haphazardly hung with dusty awards and photos, including one of Queen Elizabeth and Prince Philip and a blowup of the $4,000 cheque from the RCMP that helped get the old Waverley group started. Rows of green plastic chairs filled the room. Many of the men this night wore ball caps and woods vests with badges, and there was a handful of women, even a baby or two. There was a broad range of ages in the group, and the mood was decidedly upbeat and friendly.

"Congratulations," Rodgers boomed as he opened the meeting a few minutes late. Using the search map and the pile of documentation produced by the Whites Lake search, Rodgers reviewed every detail. He especially singled out someone's offbeat suggestion during the search that they should check the moving van that had delivered the uncle's furniture to the house earlier that Saturday. As it turned out, the van had departed before the kids even arrived — but it was an excellent suggestion anyway, since the kids might have climbed aboard and been locked in accidentally. "That's a good example of thinking outside the box," Rodgers said. "Don't get tunnel vision." There were few comments from the audience, though one or two minor improvements were considered for follow-up. Significantly, the post-operation discussion made almost no reference to technology. Rather, success was due almost entirely to the precise organization of people in the pursuit of clues. "This became a classic-type case," Rodgers said proudly at the end of the evening. "Everything was textbook."

The broken hull of MV *Flare,* which drifted for several days before sinking. TSB

CHAPTER TWO

Disasters at Sea

The Sinking of the MV *Flare*

The supply ship *Seaforth Highlander* rolled side-to-side like a giant bathtub toy, its port side blasted by southwesterly winds reaching up to 150 kilometres an hour. Swells the size of six-storey buildings flooded its afterdeck as the crew battled a howling North Atlantic gale to move a mere ten kilometres. "The wind just cut through you," second mate Jerry Higdon later told investigators. "Spray was coming over the deck, and it would freeze before it hit you. The deck was almost completely under water. It was quite dangerous to be on deck that night." The crew readied a large cargo net for a possible rescue attempt, but it was washed overboard before it could be used. The swamping seas and driving snow severely reduced visibility from the bridge. But after more than an hour of struggling in the dark, the crew could see the lights of the *Ocean Ranger* oil rig twinkling like a Christmas tree almost a kilometre to the north. The massive rig, which had sent a mayday about an hour earlier at 1:09 a.m., was listing uncontrollably, though the *Seaforth Highlander*'s first mate, Rolf Jorgensen, later said the crew could not see just how far over it was tilting. The ship could not contact anyone on the rig by radio, but *Ocean Ranger*'s last-ever transmission at 1:30 a.m. indicated that everyone was heading for the lifeboat stations. The rig that loomed ahead in the darkness was therefore likely abandoned, eerily glowing through the storm. The *Seaforth Highlander* now came across smoke flares and ten to twenty lights scattered across the surface of the heaving seas. But as the vessel drew closer, the crew saw only empty lifejackets and their seawater-activated light beacons. At about 2:14 a.m., the ship's master, Captain Ronald Duncan, caught sight of the arc of a red distress flare on his starboard side. Soon a second flare appeared in the same area, illuminating a covered orange lifeboat from the *Ocean Ranger*. At least some survivors had made it safely off the rig.

Duncan now ordered the *Seaforth Highlander* to head toward the sluggish lifeboat, which was sitting heavily in the water about 350 metres away, its bow turned into the pounding waves. As the ship pulled closer, crew members on the afterdeck could see holes punched into the bow of the lifeboat and men bailing water furiously. But the little craft was obviously under engine power because it moved deliberately around the stern of the *Seaforth Highlander* and along the port side, ready for a transfer of the men inside. The Norwegian-built eight-metre lifeboat was about a tenth the size of the big supply boat and bobbed awkwardly well below the *Seaforth Highlander*'s gunwales. There would be no simple way to get the survivors across the dangerous two-metre gap between the vessels.

The screaming wind made it impossible even to shout at the men on the lifeboat, and there were no radio communications, so both crews let their actions speak for them. Two ropes with life rings attached were thrown across to the lifeboat and both ends were made fast to each vessel, linking them tenuously. There were about thirty survivors huddled inside the lifeboat, and soon seven or eight of them emerged, inexplicably, on the side farthest from the *Seaforth Highlander*. All were wearing hard hats, and some wore only light clothing and work vests. Whether it was their sudden combined weight on the port side or some other factor, the lifeboat rolled slowly sideways away from the *Seaforth Highlander*, snapping the two lines and capsizing in the heaving seas. The men who had exited were thrown into the sea, their lifejacket lights now glowing, while the rest of the survivors were trapped upside down in the inverted lifeboat, its exposed propeller now idle. The craft was designed to be self-righting in the event of a capsizing but had obviously been so damaged — likely in a hasty and difficult launch from the rig — that it remained on its back. Some of the men hung on to the upturned craft for a few minutes, but, their limbs paralyzing rapidly in the freezing water, they soon dropped off. The crew of the *Seaforth Highlander*, meanwhile, launched an inflatable life raft and threw lines to the bobbing survivors, but their frozen arms and hands had no strength to grasp anything. Captain Duncan ordered the *Seaforth Highlander*'s high-riding propeller shut down lest it tear into any of the victims in the water, and the supply ship began to drift away from the capsized boat. Jorgensen tried to grab one man who washed up along the port side of the *Seaforth Highlander* but lost him.

Perhaps ten minutes passed between the rendezvous of the two vessels and the capsizing. The chaotic, sea-battered efforts at rescue had come excruciatingly close to pulling at least some victims from the roiling ocean, but the dismal luck of the *Ocean Ranger* that bleak night of February 15, 1982, would allow no survivors. Later that same day, at about seven a.m., another supply vessel spotted the capsized lifeboat with a *Seaforth Highlander* life ring and snapped line still firmly attached. Inside, the crew could see about twenty corpses still strapped into their positions — though several bodies floated through one of the two holes in the bow and at least one of them had washed up onto the afterdeck. The body of Melvin Fried, an employee of Odeco Drilling, was the only one of the entire eighty-four-man *Ocean Ranger* crew recovered that grim day; twenty-one more bodies were collected over the next five days. The ill-fated lifeboat is believed to have sunk after efforts to snare it failed, taking some of the sixty-two unrecovered bodies to the ocean floor, where the *Ocean Ranger* itself lay upside down like a dead animal, its two pontoons not far from the ocean surface. Somehow, every single piece of rescue technology failed that horrifying night, Canada's worst marine disaster in forty years. The only mercy was that the sea was so frigid that death from hypothermia and drowning would have come quickly.

The *Ocean Ranger* disaster, over the rich subsea oilfields almost 300 kilometres east of St. John's, Newfoundland, should never have happened. A federal-provincial Royal Commission into the tragedy cited a long list of design and safety lapses, including a glass-covered porthole that was not strong enough to withstand the typical battering of a North Atlantic storm. The gale had smashed the glass, allowing water to pour into a ballast-control station whose unprotected circuits quickly shorted out. The list that subsequently developed was made worse by the ill-advised restoration of power to the circuits, which inadvertently caused uncontrolled flooding. The poorly trained crew reacted so incompetently and so slowly, apparently unaware of the potential danger, that the situation was soon out of control, forcing abandonment of the rig. Orderly evacuation would have been nearly impossible, though there is no information — and no survivor — to describe the abandonment. "Confusion may have developed in the rush to the lifesaving equipment, and that may explain the light clothing on some of the bodies that were either recovered or sighted," said the Royal Commission's final report.

The commissioners laid all of the blame for the loss of the *Ocean Ranger* on the designers and the managers, who did not properly train the crew or supply them with the necessary manuals. "The disaster could have been avoided by relatively minor modifications to the design of the rig and its systems, and it should, in any event, have been prevented by competent and informed action by those on board," their report concluded. "Because of inadequate training and lack of manuals and technical information, the crew failed to interrupt the fatal chain of events which led to the eventual loss of the *Ocean Ranger*." The report went on to severely criticize the safety mechanisms in place to deal with such a disaster. Duncan's supply ship, which doubled as a standby rescue vessel, had none of the rescue equipment common in similar ships operating among the North Sea oil rigs, including a crane with a basket or net to haul survivors out of the water. Its very design prevented the mobility and sea access a rescue ship needs to attend to survivors in the water. Nor was Duncan provided with clear instructions on how close he was to stand by the rig; the commissioners nevertheless criticized him for being too far away that night. The *Seaforth Highlander*'s crew also had inferior clothing and training for major rescues. The report noted as well that lifeboats launched from the severely listing rig were prone to damage because they were likely to smash against the tilted structure, another design failure. The final report also found fault with the three Sikorsky S-61 helicopters under contract with Mobil to service the cluster of three rigs operating in the Hibernia oil field that included the *Ocean Ranger*. None of the choppers had hoists and other basic rescue equipment, even though, like the *Seaforth Highlander*, they were supposed to provide emergency response. "The mystique of unsinkability inhibited the kind of planning that was clearly necessary," the investigation concluded. As with the *Titanic* tragedy, which had unfolded seventy years earlier in nearby waters, hubris had blinded the operators to the necessity of basic safety precautions.

The commissioners also singled out the military's search-and-rescue Labrador helicopters for special criticism. Three of these twin-rotor choppers were based at Gander, Newfoundland, and, once refuelled in St. John's, could get to the rig in about seventy minutes. But the turbine-powered aircraft could not get airborne immediately that night because "rime icing" had been forecast for the low cloud cover between Gander and St. John's. Helicopters must avoid any possibility of ice buildup on the rotor blades, engine intakes and fuselage because the combination of added weight and changed lift characteristics can make the aircraft unacceptably sluggish and prone to crashing.

The Royal Commission's detailed analysis of the disaster timeline demonstrated that even had the Labradors been able to fly that night, they would not have made it to the rescue scene in time to make any difference. But investigators nevertheless were surprised to learn that the Labradors, acquired two decades earlier in the 1960s, were particularly ill-suited to marine rescues off the Atlantic coast. No longer being manufactured by Boeing, spare parts for the aircraft were extremely difficult to obtain, and maintenance kept them in the shop for long periods. They also lacked radar, had no standard marine radio to talk with distressed ships, and had no automatic hover system to help the pilot negotiate heaving seas while standing by a ship or hauling up survivors by hoist. Flying the Labradors was a nerve-wracking, white-knuckle operation in bad weather, and the aircraft's lack of basic technology complicated rescues unnecessarily. But the biggest fault of the Labradors, three more of which were stationed at Canadian Forces Base Summerside, in Prince Edward Island, was their relatively short operating range, barely 400 kilometres before they had to turn back or risk running out of fuel. "The main weakness of the Labrador . . . is its relatively short range and consequent lack of endurance for rescue missions offshore," the Royal Commission report concluded in 1984, more than two years after the *Ocean Ranger* disaster. Among the report's many urgent safety recommendations was that the government of Canada upgrade the existing Labradors and "obtain others capable of longer ranges and with endurance for rescue missions offshore"; the Labrador is "unsuited for marine rescue operations offshore." Specifically, the report called on the government to supply at least three long-range helicopters to provide coverage for any offshore-oil operations in the Labrador Sea and for general rescues across Newfoundland's Grand Banks and the Scotian Shelf south of Nova Scotia.

The final report on the *Ocean Ranger* disaster had a profound impact on safety for the offshore-oil sector and on the world of Canadian search and rescue. Equipment on rigs and oil-company rescue vessels and helicopters was upgraded, crew training was extended and intensified, and rig design was improved. The rescue centres in Halifax and St. John's developed detailed marine-disaster plans for the first time and obtained vital information about offshore-oil facilities to keep at hand in case of another catastrophe. One of the report's main recommendations was the establishment of a single federal agency responsible for the co-ordination of all federal search-and-rescue resources. The National Search and Rescue Secretariat came into being in 1986 and was given an administrative home in the Defence Department. The secretariat has since been active in promoting and funding basic research, in developing rescue databases, in carrying out statistical studies, and in drafting federal policies. Despite all of these advances, one of the central recommendations arising out of the *Ocean Ranger* tragedy was never carried out: the ice-vulnerable, outdated, short-range Labradors continued to provide the sole primary search-and-rescue helicopter service for all of the east coast and, indeed, for the rest of Canada. Although some of their equipment was upgraded, including the addition of radar and marine radio, they still lacked auto-hover capability, de-icing technology and the rescue range required for difficult offshore emergencies.

The chilling finality of all hands disappearing in a major marine disaster during an Atlantic winter storm was repeated four more times in the early 1990s. On the night of January 19, 1990, the crew of the bulk carrier *Charlie* passed the Italian oil tanker *Chippewa*, sending a radio message at about 10:10 p.m. The ships communicated for three minutes, exchanging navigation information, during which *Charlie* made no indication it was experiencing any difficulties despite a raging storm. It was the last anyone heard from the 148-metre bulk carrier, which had departed from Montreal five days earlier with a load of Canadian wheat destined for Sudan and Mozambique. (The 18,500 tonnes of grain was foreign aid from the Canadian International Development Agency and the Canadian Food Bank of Winnipeg.) When nothing was received for four more days — not even a distress signal — the owners reported *Charlie* and its crew of twenty-seven missing at sea somewhere just northwest of the Azores. This was at the extreme edge of the Canadian zone of responsibility for search and rescue, and so a four-engine Aurora patrol aircraft was dispatched from Canadian Forces Base Greenwood, Nova Scotia, on a 4,000-kilometre transit to the search area. A Spanish military Hercules aircraft also traced the route from the Azores to Ceuta, Spain, where the *Charlie* had been scheduled to make a refuelling stop on its way to the Suez Canal. The search was called off five days later when no trace of the Cypriot-registered ship could be found. "Can't say for sure that it sank, can't say for sure that it didn't," said a Canadian coast guard officer at the time. There has been no sign of the vessel, its Polish captain and its Filipino crew since. An investigation later found irregularities in the way the wheat was loaded at the port of Montreal.

Less than a year later, the Singapore-registered bulk carrier *Protektor* sank about 420 kilometres east of Newfoundland in another killer storm, taking thirty-three lives. The ship had been travelling to Sweden from Port-Cartier, Quebec, when it got caught in a region of pounding, five-storey seas on January 11, 1991, and radioed for help. The crew — thirty

East Indians, a Filipino, a Pakistani and one man from Ghana — huddled in the wheelhouse, awaiting rescue as the seas battered the vessel. Among them were the captain's wife and daughter. An Aurora arrived on the scene and made radio contact, only to watch the 240-metre ship disappear from the radar screen without a trace. "He was there one minute, the next minute he was gone," said Captain Reid McBride, mission commander aboard the big four-engine aircraft. "It's quite disheartening to be that close." The search, drawing on more than ten ships and other aircraft, turned up no bodies and no wreckage that could be definitively linked to the vessel — just a couple of oil slicks, some floating crates and a life raft briefly sighted. "Considering the speed at which this disaster occurred, I doubt very much whether one had time or sufficiently decent weather to make any kind of an orderly evacuation from the ship," said a coast guard official at the Halifax rescue co-ordination centre.

Two years later, there was a similar tragedy as an intense weather system swallowed a giant gypsum boat before anyone could get off safely. The bulk carrier *Gold Bond Conveyor*, with thirty-three crew on board, sailed out of Halifax harbour on March 13, 1993, into the teeth of an "extra-tropical cyclone" — weather forecasters' jargon for a monster storm that churned waves as high as thirty metres. Just why the thirty-nine-year-old captain, Man Hoi Chan, of Hong Kong, took his 177-metre vessel into such tempestuous waters is a mystery. Investigators later theorized that his grasp of English was weak, leaving him unable to fully comprehend forecasts about the magnitude of the storm. There were also questions about whether he felt compelled to haul anchor lest company profits were jeopardized, though the ship owners denied the accusations. Whatever the reason, the *Gold Bond Conveyor* and its 25,000-tonne cargo of gypsum bound for Tampa, Florida, was listing a day later about 200 kilometres off the southwestern tip of Nova Scotia. The captain's radioed trouble report, at about four p.m., prompted the Halifax rescue centre to order an Aurora into the air to monitor the ship's difficulties. At 10:37 p.m., the vessel was rolling heavily and waves were breaking over the deck. A plan to evacuate half the crew by Labrador helicopter was scuttled because of the icing conditions and winds blasting up to 130 kilometres an hour. Finally, the captain issued a mayday at 12:23 a.m. saying the crew was abandoning ship. A mere eight minutes later the *Gold Bond Conveyor* sank, the dramatic moment caught on video by the Aurora using a low-light forward-looking infrared camera. The much-televised tape shows a giant thirty-metre wave washing over the bow, running as far back as the bridge and finishing off the vessel — a rare glimpse of a major sinking. There was no imagery of anyone getting off the vessel into the nine-degree-Celsius water. A Canadian search-and-rescue technician was eventually lowered into the ocean at about noon to recover the oil-soaked body of the radio operator, Fan Chung Kong, a lifejacket having kept him afloat. "He was just wearing runners and jeans," said Corporal David Knubley, who had the grim task of retrieval. "He was probably gone within an easy half hour at the most." No other bodies were sighted or recovered, though search aircraft spotted a smashed lifeboat, two empty life rafts and a lifejacket. Nine months later the body of third mate Xiao Hui Zhang washed up off Dingle, Ireland, carried across the ocean by the Gulf Stream. The thirty-one other victims — from Hong Kong, China and Taiwan — were never found.

The extraordinary videotape and black-and-white photographs taken by the Aurora offered a unique opportunity to analyze the cause of the disaster. Using computer processing, three investigators with the Transportation Safety Board of Canada enhanced and magnified the images to show that a pair of cargo-loading doors on the starboard side of the stern had been inadvertently left open throughout the ship's struggle. The conveyor belt, used to load gypsum ore quickly, could also be seen dragging in the ocean, leaving a discernible wake. There were no automatic sensors to alert the *Gold Bond Conveyor*'s bridge to any such problem. The board's analysis said water likely flowed into the vessel through the open rear doors, sloshing through the holds and collecting in the bow, causing the growing forward list. The videotape showed the list was so pronounced that the propeller actually left the water at points. The investigators also demonstrated that the crew would not have been able to see the problem from the bridge and were likely unaware of just why they were taking in water, perhaps thinking the problem was in the bow rather than the stern.

Nine months later, another bulk carrier disappeared with all hands during a fierce Atlantic storm. The Liberian-registered *Marika 7* left Sept-Îles, Quebec, on December 27, 1993, with 150,000 tonnes of iron ore bound for Ymuiden, Holland. Five days later, on New Year's Day, Canadian rescue officials received a satellite-relayed distress signal from the ship's floating emergency beacon, which is triggered automatically when it hits seawater. There had been no radioed mayday or any earlier indication that the 295-metre ship was in trouble, and all attempts to contact the thirty-six crew members were met with silence. The beacon's signal indicated the ship was almost 1,500 kilometres east of Newfoundland, well inside the Gulf Stream, where water temperatures were about fifteen degrees Celsius. The Halifax rescue centre sent a four-engine Hercules aircraft from Greenwood, Nova Scotia, to begin the search through the soupy mix of rain, fog and snow; winds gusted up to sixty kilometres an hour. Once on scene, the crew reported seeing two green flares but no sign of the *Marika 7*. The Hercules and several commercial ships diverted to the area also reported seeing lights in the water, as well as two life rafts, a broken life ring, oil slicks and debris. "The fact that no distress call was received from the *Marika 7*, even though she carried modern communication equipment, suggests disaster occurred suddenly and without warning," says a Canadian report on the sinking. "This would likely result in a less than ideal attempt at abandonment." No other flares were sighted during the next five days, suggesting that, if some of the crew had made it safely to a life raft, they did not survive long. The two self-inflating life rafts that were found (the ship's third life raft was never sighted) showed no signs of ever having been occupied. Anyone in the water was likely dead from hypothermia within five hours despite the relatively balmy seas. None of the bodies of the six Greek officers and thirty Filipino crew members was ever recovered.

The Canadian military aircraft assigned to the search for any *Marika 7* survivors were operating at the far edge of the Canadian zone of responsibility, beyond the Grand Banks and Flemish Cap. The post-mortem report on the operation noted that none of the aircraft was equipped with Global Positioning System (GPS) navigation hardware, which uses American military satellites to determine precise location on the earth's surface. The commercial ships diverted to aid the *Marika 7*, however, were equipped with GPS, making it difficult for the military pilots to direct the vessels, since they were operating with navigational errors based on their older cockpit technology.

The most serious technological drawback in the *Marika 7* rescue mission — and indeed in the *Charlie*, *Protektor* and *Gold Bond Conveyor* disasters — was the inability of the Labrador helicopters to take on any meaningful role. Big bulk carriers can develop major problems in the space of mere minutes, and even benign sea temperatures can kill men in the water in a few hours. Helicopters, the most versatile rescue platform because of their ability to hover and hoist, are highly effective life savers if only they can be brought into play. But the twin-rotor Labradors sat uselessly on the tarmac or in their hangars, either because they lacked the range to operate at the great distances involved in the *Charlie*, *Protektor* and *Marika 7* disasters, or because they could not fly in severe icing conditions during the *Gold Bond Conveyor* operations. These were the very problems cited by the *Ocean Ranger* report, issued a decade before the *Marika 7* sinking. Despite the urgent appeal to acquire long-range choppers with more all-weather capability, the aging Labradors were entering their fourth decade of service with no replacements on the horizon.

Later in the same year as the *Marika 7* sinking, another marine disaster on Canada's doorstep dramatically illustrated the tactical advantages of helicopters that had rescue ranges appropriate to east coast search zones, which stretch to the middle of the Atlantic. The cargo ship *Salvador Allende* was traveling from Freeport, Texas, to Helsinki, Finland, with a load of rice when it ran into a storm some 1,500 kilometres off the south coast of Nova Scotia. The 150-metre vessel was apparently hit during the night of December 8-9, 1994, by two "rogue waves" that shifted its cargo, causing a pronounced port list of forty-five degrees and more. The tilt also shifted the lubricating oil in the ship's engines, causing them to shut down. The engine-room crew managed to restart them, but the captain simply could not manoeuvre the *Salvador Allende* to a heading that would at least stabilize the list. Six hours later, with hope dwindling, the captain ordered the motorized lifeboats launched. The stern of one of these orange vessels was sheared off in the attempt, causing it to sink with no one aboard. The second lifeboat was put into the water successfully with eleven of the thirty-one crew members aboard, including two women who worked in the ship's galley. The crew also managed to launch three of the four life rafts, but the heaving seas kept breaking the lines, carrying them off before any survivors could climb aboard. As the ship began to sink rapidly, the remaining crew members went into the water, which, because they were in the Gulf Stream, was at least seventeen degrees Celsius.

The *Salvador Allende* had gone down inside the American zone of responsibility, south of the Canadian search-and-rescue region. As always in such cases, both countries co-operated in the rescue by providing resources, though the New York rescue centre was in charge of co-ordinating the operation. A Hercules aircraft from Greenwood, Nova Scotia, was first on the scene, arriving shortly before the *Salvador Allende* slipped beneath the waves. The aircrew reported that they saw life rafts and survivors in the water, and they dropped survival kits that included food, water and radios. Meanwhile, two U.S. Air National Guard helicopters from 106 Rescue Wing were sent to the scene from their Long

Island, New York, base. These were HH-60G Pave Hawks, twin-engine helicopters built by Sikorsky. Designed for combat, they were also used for search and rescue partly because they have de-icing technology to keep the engine and rotor blades free of ice in cold weather. The single-rotor Pave Hawks also have flight-control systems to allow automatic hover. But their most useful feature for search-and-rescue work is a retractable refuelling probe that allows them to replenish their gas tanks during flight by connecting to a hose trailing from a Hercules aircraft fitted as a tanker. The procedure gives them virtually unlimited range. The Long Island Pave Hawks — accompanied by three Hercules, one each from the Air National Guard, the Marine Corps and the Air Force — stopped first at the Shearwater, Nova Scotia, air base near Halifax before heading to the rescue site far out at sea.

One of the Ukrainian sailors from the sunken *Salvador Allende*, third engineer Alexander Taranov, was among a group of four people washed overboard who tried to swim together to one of the life rafts dropped by the Canadian Hercules. But he became exhausted fighting the swells and finally gave up, clinging instead to a floating pallet and losing track of his three companions. The wooden structure kept knocking him in the tossing seas, so Taranov simply let go, trusting instead to his flimsy thermal overalls, two sweaters and a lifejacket. At thirty-six, Taranov was a robust 240 pounds, and his youth and body fat greatly increased his chances of survival despite his inadequate clothing. For thirty-two hours he valiantly hung on, fighting cramps and at one point grabbing two cans of 7Up that were bobbing among the flotsam. He drank one and tucked the other inside his clothing. Taranov slept for short periods, perhaps twenty minutes at a time, before a wave would smack him awake again. The aircraft droning above kept alive the hope he would see his wife and two daughters again at their home in Kherson, Ukraine. "There was no doubt in my mind they would get me out," he said later. "I was not afraid."

The Pave Hawks continued their long ocean flight, eventually refuelling in the air ten times and setting a world record for the longest over-water helicopter rescue. The flight engineer from one of the helicopters spotted a waving survivor — Taranov — about 112 kilometres south of where the *Salvador Allende* had sunk. The crew also sighted sharks in the area, increasing the pressure to get the job done soon. A pararescue specialist jumped into the water and worked quickly to get Taranov — blind from the salt spray — hauled up by hoist, an extraordinary rescue largely due to the endurance and versatility of the Pave Hawks. They made for Shearwater again, where doctors recorded Taranov's temperature at a normal 37.8 degrees Celsius. Another survivor, the thirty-six-year-old second mate Skiba Ivan, climbed into a life ring tossed out by the *Torungen*, a passing commercial ship pressed into the search by New York. Ivan was also reported to be in good condition as the *Torungen* headed for Europe. None of the twenty-nine other crew members survived, and although some bodies were sighted, none was ever recovered. Some Canadian rescue officials believe that the numerous sharks seen in the area might have caused the early deaths of some of the survivors in the water.

Another "miracle man" survivor was plucked from the seas almost three years later after the sudden sinking of another cargo ship. On October 23, 1997, the 109-metre *Vanessa* got into trouble more than 800 kilometres east of Newfoundland while en route to Bolivar, Colombia, from Copenhagen, Denmark, with a load of fertilizer. The cargo, packed in more than 3,200 bags, shifted badly in heavy seas, creating a seventy-degree list from which the ship never recovered. A vessel tilting to that extreme reaches a point called "deck-edge immersion" at which the form of the hull no longer provides any buoyancy. The sudden emergency prevented the radio officer from making his way off the bridge to the signals room and sending out a satellite mayday, so he had to rely on a small, hand-held unit with minimum range. Luckily another vessel was in the area, the container ship *Choyang World*, which picked up the signal and relayed the mayday to rescue officials in Norfolk, Virginia. As *Vanessa* sank, nine crew members — including the captain, Roberto Barruga — made it safely into a life raft, but the remaining six were washed off the deck into the water. As with the *Salvador Allende* sinking, the survivors had the good luck to be inside the Gulf Stream, with sea temperatures reported between seventeen and twenty degrees Celsius.

The Norfolk rescue centre quickly alerted its Halifax counterpart that the *Vanessa* had gone down inside the Canadian rescue zone. The Canadian Coast Guard ship *Cape Roger* was diverted to the area, though it would take more than eight hours to arrive. A Hercules and an Aurora, both from Greenwood, were also sent to the site. A commercial ship, the *Summer Wind*, rescued the nine survivors in the life raft about nine hours after the *Vanessa* went to the bottom. The *Cape Roger* arrived on the scene about ten minutes after the *Summer Wind* rescue and began searching for the missing six crew members, all of whom had only lifejackets for survival. The crew of the coast guard ship recovered a dead body six hours later, and in another four hours thought they had come across two more bodies. A fast rescue craft was sent out from the *Cape*

Roger, and crew member Dwayne Barron was about to haul the two bodies out of the water when he got a fright — one of the figures opened his eyes and reached up, grabbing him.

Henry Almonte, thirty-four, had spent nineteen hours floating in the ocean, gripping his dead crewmate for most of that time. Wearing pants, a sweatshirt and a leather jacket underneath his lifejacket, Almonte had estimated it would take perhaps five hours for rescue, so he relaxed early on to conserve his energy as much as possible. Whenever he began trembling, he grabbed his feet and tensed to stop the shakes. His friend, the chief engineer, did not know how to swim, so Almonte comforted him by holding on tight, though the man died sometime during the ordeal. "If I cannot make it, if they don't find me, God will make his own decisions," Almonte thought to himself between prayers. Once aboard the *Cape Roger*, Almonte was found to be severely hypothermic, so a coast guard officer stripped to his underwear and got into bed with him to share body heat. The officer also kept talking and telling jokes to prevent Almonte from nodding off and making his weakened condition worse. Two search-and-rescue technicians from the Hercules also parachuted into the sea and were picked up by the *Cape Roger*'s fast-rescue craft to give first aid to this tenth and last survivor of the *Vanessa*. Almonte quickly revived, and after a short stay in hospital in St. John's he was reunited with his wife and two children in the Philippines.

Absent throughout the difficult *Vanessa* rescue operations were Canada's Labrador helicopters. Once again, the sinking occurred well beyond the limited range of these aging workhorses. One Labrador from Gander was moved to St. John's and kept on standby in case the returning *Cape Roger* came within range and required an emergency medical evacuation. But the chopper was never needed and returned to its base having remained idle throughout the *Vanessa* operation. Unlike the rescue of the "miracle man" of the *Salvador Allende* by a refuellable, weather-tolerant American Pave Hawk, there were no helicopter heroics on this mission, more than fifteen years after a royal commission urged prompt replacement of the entirely unsuitable Labradors.

■ ■ ■

Tony Isaacs and Petar Markovic first met on the slippery spine of an overturned lifeboat in a heaving, frozen sea somewhere between Cape Breton Island and Newfoundland. Tony wore an insulated wet suit and mask, and his heart was pounding hard from an arduous swim through four-metre seas. Petar was wearing street clothes and an inadequate lifejacket, soaked to the skin and suffering from moderate hypothermia. A pungent film of dark fuel oil coated the ocean and the lifeboat, making it a miserable, slick oasis. Petar and three companions were lying prone on the bobbing back of the fibreglass boat, holding on tightly and shivering constantly. They had clung like half-frozen leeches for more than five hours, trying to keep their numb limbs completely out of the water. The minus-two-degree Celsius winds and stinging spray across the back of the boat gave them only a fractionally greater chance of survival than if they were immersed.

Earlier on that wintry morning, neither Tony nor Petar had had any inkling that such an unusual rendezvous was in store. The web of fate and circumstance that drew them together was woven through hundreds of other lives and extended across thousands of kilometres, and it was rooted in a half-century of Canadian rescue culture. Each rescue is unique, yet this rescue seemed to draw together all the pieces of the Canadian rescue system — the planning, the training, the technology and the simple human valour.

The origins of that remarkable meeting are complex, but the day itself began rather simply. Both men were asleep, recovering from jobs that were intensely physical and, for the most part, routine. Isaacs slept alone in a room at the Cambridge Suites in Sydney, Nova Scotia, a waterfront hotel with a commanding view of the Sydney River as it broadens into the south arm of the harbour. Isaacs was a military man, a sergeant, based at the Greenwood air force base in Nova Scotia's Annapolis Valley. His profession: search and rescue technician, or Sartech, an elite, bone-jarring trade that exists nowhere else in the world. One of eight graduates of the Sartech class of '83, Isaacs was trained in a broad spectrum of rescue skills, from emergency medicine to scuba diving, from wilderness survival to parachuting. He could drop water pumps accurately from the ramp of a Hercules aircraft to distressed vessels. He could launch smoke cylinders to mark the precise location of life rafts. He was trained in radio, in the use of night-vision goggles, in hoisting, in administering oxygen and morphine. Indeed, to keep his broad range of skills current and sharp, he and all of the other 130 or so Sartechs in Canada have a rigid schedule of training, retraining and review.

Isaacs was on just such a training mission in mid-January 1998. He and a junior Sartech, Corporal Paul Jackman, with just six months' experience, had joined the crew of a Labrador helicopter for a three-day RON — military jargon for "remain over night." The five were to fly to

Sydney, then to Prince Edward Island and back to Greenwood over three days. Normally, search-and-rescue training is done within the space of a single workday at Greenwood, with short hops to other airports in the Atlantic region. Lunch might be spent eating bad airport food at Stephenville, Newfoundland, Summerside, Prince Edward Island, or Saint John, New Brunswick. But these RONs took crews away from base for longer trips two or three times a year. This particular RON was going rather poorly, however. The weather was snowy, with freezing rain, high winds and low cloud ceilings, and Labrador helicopters are notoriously poor performers in icing conditions. The crew had to wait out the bad weather at Sydney and were now stuck in the Cambridge Suites. Each crew member was assigned his own room, rather than economically doubling up, for a very good reason: each had to get a solid eight hours of sleep, uninterrupted by the snoring of a crewmate, to be well rested for whatever the next genuine mission might hold.

The weather had improved somewhat that Friday morning, January 16, and the team was expecting to head back to Greenwood after a disappointing training exercise. They all packed their personal kits and met at the buffet breakfast room on the main floor of the hotel for some unexceptional coffee. As they were finishing at about seven-thirty, the aircraft commander, Captain Chris Brown, got a call from the rescue co-ordination centre in Halifax, routed through the front desk. The bright yellow Labrador, tail number 304, was being assigned to a major search-and-rescue operation off the south coast of Newfoundland. Two other Labradors had already been "tasked" to the search mission (in the military vernacular), and with this third helicopter, the operation would always have two choppers in the air while the remaining one refuelled. The Sydney Labrador was now officially Rescue 304, based on its tail number. Brown was told to get the twin-rotor aircraft ready for flight, details to follow during the pre-flight check. The five climbed into their rented van and headed for the Sydney airport.

While the flight engineer, Master Corporal Rob Butler, and co-pilot Captain Richard Gough joined the Sartechs in preparing the aircraft and equipment for flight, Brown received his orders from Halifax. He was to proceed directly to the airport at St-Pierre-Miquelon, islands owned by France off the south coast of Newfoundland, where he would refuel to have the maximum time for an air search. The crew was to watch for survivors of a bulk carrier, the MV *Flare*, which had apparently foundered in rough seas, though there were scant details. There had been no sign yet of any of the twenty-five crew, dead or alive. Indeed, searchers hadn't found anything — the ship, lifeboats, life rafts, debris, oil slick, survivors — nothing at all. There were only some vague co-ordinates based on a sketchy mayday call from the ship.

With the Labrador checked and ready, the crew took off for St-Pierre-Miquelon, about an hour to the east. They weren't in the air for more than five minutes when Butler noticed some red-coloured fluid pouring out of the transmission box while he was doing an in-flight leak check. Alarmed, the pilots declared a "pan pan pan," literally a breakdown, from the French word *panne*. They immediately turned back to the airport: without fluid, helicopter transmissions soon quit, and with no transmission, the rotors get no power and a chopper must land immediately. An Air Nova Dash 8 passenger aircraft was just making its final approach for one of the Sydney Airport runways, but, with the potential emergency, the pilot elected to overshoot, clearing the way for the ailing Labrador. The crew landed safely, and once the twin rotors had been stilled, Butler carried out a closer inspection. The red-coloured fluid, it turned out, was mostly water mixed with some old transmission fluid that had accumulated over many months in an inaccessible area. The water, Butler determined, originated from the snow and freezing rain that had worked its way into the chopper during the two-day weather layover in Sydney and had then turned to ice. As the aircraft cabin warmed up during the short flight toward St-Pierre, the ice melted, triggering the false alarm. But with an aircraft that had been mission-battered for more than three decades, he simply couldn't afford to take chances.

Forty-five minutes later, the Labrador was back pounding the air, skirting some fog just off Sydney and heading for a search area the crew was asked to check as they made the transit to their refuelling stop in St-Pierre. The proposed search area was less than two kilometres off their intended track and would at least make good use of their otherwise wasted transit time. Cruising at an altitude of about 150 metres, both pilots soon spotted a huge oil slick about eighteen kilometres back from the assigned search area. Like gasoline in a street puddle, the slick had a brilliant rainbow sheen and was between ninety and 120 metres wide. Bobbing inside its dark boundaries were unidentifiable bits of debris. "Usually an oil slick is a pretty good indication of something gone wrong," Chris Brown recalled later. "We knew we had found the point where the ship had gone down." The slick — a foul mixture of 130 tonnes of marine diesel fuel, 550 tonnes of heavier fuel oil and another 13,500 litres of lubricants — stretched more than a kilometre on an east-west axis, pulled along by the wind like taffy. Brown brought the aircraft lower, following this heaving liquid highway down the middle in search of possible survivors. As his co-pilot Richard Gough radioed the sighting to a Hercules aircraft that was co-ordinating the search, they spotted an overturned, orange-coloured lifeboat at the easternmost end. The wind had grabbed the 7.5-metre vessel's freeboard, the part that rose above the water line, and driven it more rapidly than the mass of the oil toward the far end of slick. The shimmering fuel had acted like a giant pointer indicating the boat's precise location.

The Labrador crew made out some human forms on the top of the lifeboat, all of them stretched precariously across the icy keel. As the lights of the Labrador and its pounding rotors came closer, it stirred those forlorn figures to life. "They were waving like mad," says Gough. Three of them were kneeling or crouching, while a fourth was lying on his side. The two Sartechs, who had not expected to be on scene at the officially assigned search area for at least fifteen more minutes, now quickly got into their wet suits and fins and began checking equipment. The aircraft did not carry full scuba gear, the result of a fatal Labrador crash in British Columbia that was partly the result of an overloaded helicopter. Sartechs now could only carry diving gear when it was known before takeoff that scuba equipment would be needed. Isaacs and his partner would have to make do without, which meant exposure to nauseating oil and the risk of hypothermia.

Brown did three or four "orbits" of the fibreglass lifeboat, wide circles to allow the flight engineer and Sartechs time to get ready, but also to become better aware of the difficult environment in which this rescue would take place. The pilots switched to an antenna on the top of the aircraft rather than on the bottom in case they had to ditch in the ocean. Ditching was to be avoided in the Labrador. Although the aircraft is designed to float on its belly in an emergency, the heaving seas would knock it over and sink it in a few minutes. On one of these slow orbits of the lifeboat, the pilots spotted a second lifeboat-turned-turtle about 100 metres from the first but with no survivors clinging to its back. Co-pilot Richard Gough now radioed in a "notice of crash location" or NOCL, a short message that announced they had found the site of the disaster. (The reference to "crash" was borrowed from rescue procedures as originally applied to air disasters.) A standard NOCL also includes the precise location and colour-coded status of the missing persons. Four blacks, for example, means four dead bodies. But with these four still to be rescued, the message read only "undetermined," or medical status unknown.

As Brown carried out his meticulous reconnaissance, he was careful not to come too close to the four half-frozen men. A loaded Labrador helicopter weighs about ten tonnes, and in a low, slow hover, its two rotors must buoy that mass by continuously displacing ten tonnes of air that slam hard against the surface of the earth, whether it's land or water. The underside of a hovering Labrador can be an extremely violent place. Because it has two rotors, the Labrador has a peanut-shaped "footprint" with an area of somewhat lesser turbulence in the middle of the peanut where the hoisting of victims can take place in relative calm. But as the aircraft moves into position, its rotors create a dangerous cyclonic downdraft, known as a rotor wash, as the spinning blades cross over a boat or victim. Accordingly, Brown kept his aircraft far enough away from the four exhausted men so they wouldn't be knocked off their pathetic boat into the oily, churning sea. At least the flying conditions themselves were positive. The cold air, at minus two degrees Celsius, was denser than warm air, so the Labrador's twin rotors did not have to work so hard to give the aircraft its lift. In addition, the rescue would take place at sea level, where the air is always denser than at higher altitudes.

Tony Isaacs, the veteran Sartech, was now ready. The plan was for Isaacs to be lowered down to the boat to get all the men into the Labrador in four separate hoistings. His Sartech partner, Paul Jackman, would help with first aid and be a backup in case anything went wrong with any hoisting. Isaacs would use a rescue collar, shaped like a horse collar, that has become standard kit for Canadian rescues. The U.S. Coast Guard prefers using a basket that is lowered from a helicopter without a rescue technician; survivors must somehow get themselves into the basket. But the Canadian military, through long experience, has settled on a system in which a Sartech assists the survivor by being lowered in a harness for himself with a rescue collar for the victim. The survivor holds himself in the collar, which goes around his upper chest and under his armpits, by keeping his arms lowered. Both Sartech and survivor are then hoisted like Siamese twins. The Canadian system is considered faster and more certain, though there is a theoretical risk. Some medical authorities suggest that hypothermia victims should remain horizontal at all times, a position that is impossible with the rescue collar but can be accomplished with the U.S. Coast Guard basket. The Canadian response to the medical concern is that the supposed risk is worth taking, given the importance of speed and the fact that victims are often weak and disoriented and need on-the-scene help.

Rob Butler, the flight engineer, was to act as the mission's hoist operator, an unglamorous but exacting job. The Labrador has two hoists, one at the right front hatch, immediately behind the aircraft commander and separated by an internal wall. This is the main, preferred hoisting position. A second hoist is available in the centre of the aircraft, but the mechanism itself is slower, and the flight engineer has to lie on his stomach to operate it. The side hoist is much faster, and, because it is located closer to the main pilot, he can better gauge his position relative to the point of rescue. This crew, however, was taking a slight chance with the side hoist. Before they had left Greenwood, the mechanism was declared unserviceable because of an apparently faulty part. Any hoisting during their training missions would have to be done with the less desirable centre hoist, they were told. But checks Butler did in Sydney had indicated that the problem was overstated and perhaps non-existent. "We elected to work with the front one until it packed it in," Butler recalled. As luck would have it, the side hoist never did quit.

Tony Isaacs was finally lowered twenty metres in his harness into the sea, splashing down about seven metres from the lifeboat. This was deliberate, to keep the violent rotor wash away from the survivors. Isaacs had to swim through waves cresting as high as five metres, climb aboard the greasy lifeboat and help Markovic quickly into the rescue collar. The swim was so arduous that Isaacs realized immediately he would not have the stamina to hoist all four survivors; he would look after the first two, and Paul Jackman would have to pick up the last two. Back aboard the Labrador, meanwhile, Butler had to play out the hoist line to give Isaacs enough slack to move around but not enough to allow the line itself to become a hazard that could wrap around a leg as the boat lurched up and down. This in fact happened once during the rescue, but Butler was quick enough to reel out sufficient line to avoid yanking a man into the sea.

Pilot Chris Brown was holding the Labrador in a steady hover, using the lifeboat as a reference point at his two o'clock position. The Labrador's lack of automatic-hover technology puts enormous pressure on pilots during such operations. Over the ocean especially, it is difficult to be certain just how high you are and whether you are keeping the aircraft stationary. The rotor wash can create concentric circles of waves moving outward, giving the illusion of descending into a bottomless pit. A pilot must constantly resist the impulse to increase altitude. In addition, the wind-driven waves may create an illusion of linear movement where none exists. Fortunately, this rescue was not taking place at night when these distracting phenomena are compounded. Co-pilot Richard Gough continuously called out altitudes — "conning," as it's termed — but Brown still needed a reference point, such as the bobbing lifeboat, to help keep the Labrador stable. Now that Isaacs had the collar around Markovic and was signalling the all-clear for hoisting, Brown had a problem. He had to bring the helicopter directly over the lifeboat, losing sight of his only reference point. Butler would help ease him there by calling out directions, as in "move forward and right ten units," with "units" being an arbitrary measure used only as a rough guideline for this manoeuvre, and Gough kept calling out the altitude. Brown, with neither instruments nor visuals, thus relied entirely on the advice of his crewmates to get him safely over the lifeboat.

With the Labrador now directly above Isaacs and Markovic, Butler quickly reeled them in like a pair of fish. They were helped awkwardly through the narrow side hatch onto the floor of the cabin. Markovic's eyes looked cloudy, and he was mumbling incoherently — sure signs of advancing hypothermia. Isaacs prepared for another exhausting descent, but before being lowered he told his Sartech partner, "Get ready, because you're going after this." Brown had by now pulled the Labrador back and left, so he now had the lifeboat for reference again at his two o'clock position and was holding steady. Down went Isaacs into the sea for another gruelling seven-metre swim to the slick back of the boat to get a second man into the harness. Soon he returned with Cyrus Ferraren, a young Filipino who was severely hypothermic, far worse than Markovic. Now it was Jackman's turn to descend into the oily swell. As he got a third man, Peter Soriano, into the harness, the last fellow on the life raft — deranged and panicky — managed to grab Jackman's leg. Jackman was able to shake him loose, promising he'd be back. "I'm used to dealing with people who are much more coherent, but these people were just like lifting sacks," Jackman recalled. "Their limbs just weren't working, so you literally had to lift them up." With Soriano now safely in the cabin of the crowded Labrador, Jackman returned for the last survivor, Remarlo Napa, who was clinging for dear life to an ice-covered rope on the life-boat.

Napa was in the worst shape of all four. He was severely hypothermic. It was a medical marvel that he could move at all, yet he had already interfered with one rescue and was now proving to be a handful as he tried to cling blindly to anything near him. Jackman managed to get him into the Labrador, but then Napa started to go berserk, thrashing about. "He had this really bewildered look, looking right through us. He just kept grabbing at everything," Jackman recalled. As Butler later described the reaction, "Everything shuts down but the animal instinct." Isaacs wound up having to subdue him physically in the highly confined space of the Labrador cabin. "He had a great will to live," Isaacs said. "He had a T-shirt on and a pair of underpants — nothing else. And in his own mind he said, 'As long as I hold on to this rope, I'll be safe.' And that's all he did. And when Paul took him off, he was still gripping the rope, even when we pulled him off. So when he got into the helicopter, he was fighting because he wasn't gripping anything. And any time he grabbed hold of something, he wouldn't let go."

At one point, Napa grabbed the black intercom cord used by the cabin crew to communicate during flight. Isaacs finally decided to sit on top of Napa to stop his dangerous flailing. "He was lying on the floor in the fetal position, and he stood up with me on his back like I wasn't even there. I was very impressed with his strength," Isaacs said later. As Brown piloted the Labrador toward the nearest hospital in St-Pierre, Jackman used scissors to quickly cut the cold, dripping, greasy clothes off the four survivors. The scissors were necessary not only for speed; there was also a danger of sending the frigid, pooled blood in the limbs up to the vital organs by trying to pull tight shorts and shirts off arms and legs. The injection of cold blood into the heart tissue and chambers can trigger ventricular fibrillation and heart failure, risking immediate death by a medical phenomenon known as afterdrop.

The four survivors were given warm winter clothing, and their vital signs were checked — blood pressure, temperature, pulse — while they were offered hot fluids. One man's feet were so black they looked severely frostbitten, but they were only covered in thick oil. The oil made the legs of the others look as if they were wearing nylons. Butler and Jackman could count on no help from Isaacs, who was reduced to being a dead weight on top of the combative Remarlo Napa. "He was the cook," Isaacs said later, "Maybe he got some extra rations." Markovic, now much recovered, began to demand cigarettes to calm his nerves. He was politely but firmly refused. During the flight to St-Pierre, seventy-five kilometres to the northeast, Brown made a short diversion to check an inflated rubber life raft with a covered top that they spotted along the route. This lighter vessel with even more freeboard had been blown about 500 metres farther downwind than the lifeboat. The Labrador made a slow, creeping pass, avoiding a stationary hover that might knock over the rubber craft with rotor wash. As far as the Labrador crew could determine from peering into its tent-like entrance, the life raft had no one inside. Another Labrador crew arrived later to lower a Sartech, who deflated the covered vessel by stabbing it with his knife and sinking it. Search crews would not uselessly check the boat again.

Thirty minutes later, Brown's Labrador set down at the St-Pierre airport, where it was met by ambulances, gendarmes and French medical officials, who quickly whisked the four survivors away to hospital. "Too much cold sea, too much cold," Markovic said from his stretcher. One of the four, Peter Soriano, registered a core body temperature of just 26.8 degrees Celsius — more than ten degrees below the human body's natural level of thirty-seven degrees. A core temperature below thirty-two degrees is considered dangerously hypothermic, when unconsciousness and death are imminent. It was a miracle these men were still alive. Markovic, who had never been to sea before, told visitors at his bedside that he had been prepared to slit his own throat with his pocket knife rather than die in the cold sea. Napa said he couldn't remember a thing about his ordeal.

The firefighters at the airport let the Sartechs use their industrial-size washing machines to get the oil out of their clothing while everybody got a chance to eat. Two hours later, the crew was back in the air determining whether the floating, drifting bow of the *Flare* had its inflatable life raft still in place on deck. It did, between the No. 1 and No. 2 holds, eliminating the need to search for it. The played-out crew returned to St-Pierre afterward, turned their aircraft over to a fresh crew and waited for two days for the weather to clear before hitching a ride on a Hercules back to Greenwood and their families. In the meantime, they visited their four former passengers in hospital — an extremely rare occurrence for Canada's military rescue crews, who often never hear again from the people they pluck from disaster. At the bedsides of Petar, Remarlo, Cyrus and Peter, they learned all about a very bad morning for the sailors aboard the MV *Flare*.

■ ■ ■

Petar Markovic's day began after waking unexpectedly from a well-deserved rest in a cabin at the stern of the *Flare*, which was sailing west through bad weather along the south coast of Newfoundland toward the St. Lawrence River. This voyage aboard the 16,398-tonne bulk carrier had begun in Rotterdam on December 30, 1997, when the empty ship — carrying only light ballast — headed across the North Atlantic to pick up a load of wheat in Montreal for the Sudan. The 181-metre vessel had been built at the Hakodate shipyard in Muroran, Japan, in November 1972, making it one of the oldest bulk carriers still plying the North Atlantic. (Until 1987, it was known as the *Doric Flame* and until 1989 as simply *Flame*.) One of some 3,000 bulk carriers around the world, *Flare* was in the smallest "Handysize" class. Notwithstanding its relatively advanced age, *Flare* had been given a major three-month overhaul beginning in November 1995, in Shanghai, China, and it was equipped with relatively modern rescue equipment. *Flare*'s hull also received extensive steel renewal during the 1995-1996 overhaul, when the seven 14.5-metre-tall cargo holds were strengthened for hauling ore. The hull was black, and the ship had a white five-storey superstructure at the stern containing the bridge and living quarters. The diesel engine, powering a single screw, and the funnel took up most of the room at the stern. The ship was capable of traveling at about sixteen and a half knots, or more than thirty kilometres an hour, but typically went no faster than twelve knots, or about twenty-two kilometres an hour. A radar mast soared forty metres above the keel, and a radio antenna about thirty-five metres tall stood at the front of the bow. Six big cranes ranged across the broad red deck. Crew quarters were located on the outer perimeters of the stern.

Flare was inspected again in November 1997, less than three months before the sinking. This survey, by Lloyd's officials, was carried out at the port of Cienfuegos, Cuba, and it revealed some metal corrosion. But the problems were not considered an immediate safety hazard. The ship was allowed to sail with the requirement that repairs be undertaken before the end of February 1998. All of the necessary Lloyd's certificates — including those attesting to the health and competency of the captain, officers and crew — were complete and up to date. Registered to owner ABTA Shipping Co. Ltd. in Limassol, Cyprus — a flag of convenience — it carried a crew of twenty-five: three of them Greek, three Yugoslavian, two Romanian and the rest Filipino. This would be the captain's first command of a bulk carrier and his first voyage aboard the *Flare*, though he had served as a first mate on similar vessels. The ship was equipped with a constellation of high-tech safety and rescue equipment based on new international standards, known as the Global Maritime Distress and Safety System or GMDSS. These standards were to come into universal effect February 1, 1999, but *Flare*, like many other modern cargo ships bigger than 300 tons, had them installed well before the deadline. Lloyd's does not set any age limits for the vessels it insures but does require more thorough inspections the longer a ship is in service. *Flare*

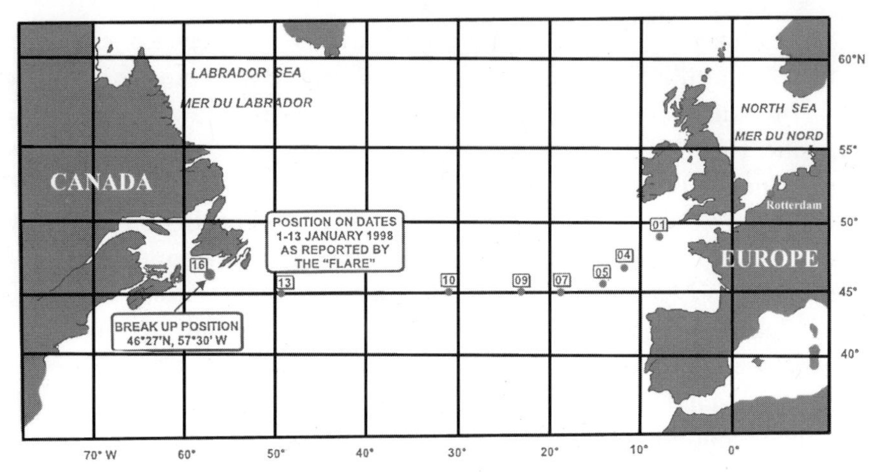

A chart of the *Flare's* crossing of the North Atlantic in January 1998. TSB

had passed these tougher requirements. On paper, at least, it looked like a safe ship.

The view from the deck was far more worrisome, especially to the men who had to sail *Flare* across the stormy North Atlantic. Filipinos who had been recruited in Manila to join the ship in Rotterdam were warned by the departing crew that the vessel was too rusty and old to make another North Atlantic voyage safely. The warning was contrary to assurances they had received in the Philippines that *Flare* had been well maintained. "Crew members signing off were telling us about the poor condition of the ship," Remarlo Napa recalled later to a CBC interviewer. "They thought it would not make it through the voyage." These crewing contracts, however, were a much-sought-after ticket out of poverty in the Philippines. As Peter Soriano put it, "I have no choice because I need money also. In the Philippines, I have no job. It's better to go on board." The recruits simply swallowed their fears and joined the crew. Raw statistics suggested their fears were justified: in the eight-year period between 1990 and 1997, ninety-nine bulk carriers around the globe had sunk — more than one a month — killing 654 crew members. These ships had proven to be much more dangerous than the oil tankers or fishing vessels plying the same seas.

The weather was miserable throughout the *Flare's* voyage to the southeastern edge of Newfoundland. The captain ordered a course south of the great circle route, likely in an effort to avoid the worst of the weather. Two days out, as they were celebrating the New Year in the officers' mess, the ship pitched so dangerously that glasses and plates slid off the tables. One of the ship's officers went pale, confessing to the men his fear that such an extreme pitch could crack the ship in two. The crew later had to weld down the galley stove after it broke free because of the violent wrenching of the sea. Over the next few days, holes and cracks started to appear in the cavernous cargo holds as the waves rose to sixteen metres. While in Rotterdam, the rookie captain had brought aboard a portable welding unit and steel plate and bars to carry out repairs during the voyage. The crew now fixed a hole in the port topside tank in hold No. 3 and another hole in the starboard topside tank. A metre-long crack appeared in the bulkhead between the No. 6 and No. 7 holds as the heaving Atlantic twisted the lightly ballasted hull. Strong winds and waves also knocked a piece off the radar mast. The rolling made it difficult to sleep. "One of the survivors reported seeing the main deck flexing such that deck cranes appeared to be touching each other," Canadian investigators later reported. "One survivor was so apprehensive that he kept his cabin light on and practised dressing in warm clothing as quickly as possible."

On January 15, the day before the accident, the ship made mandatory contact with the Eastern Canada Traffic Regulation System or ECAREG, reporting its position and advising that sea conditions were so bad that the captain was reducing speed to three knots, or less than six kilometres an hour, as it crept west about seventy-five kilometres from shore. "Very bad weather" read the short radio transmission, which estimated the ship's arrival off Cape Ray, at the southwest tip of Newfoundland, sometime the following morning.

That night conditions grew much worse. Waves rose to seven metres, and the wind — coming from the southwest — increased to about seventy-five kilometres an hour. At midnight, there was a loud, slamming bang and the entire hull quivered like a saw blade. Shortly before five a.m. Newfoundland time, another thunderous bang and shudder ran through the eerily vibrating ship. The vessel had actually snapped in two, in its middle, like a piece of chalk. The general alarm sounded continuously, and crew members jumped from their beds in panic as *Flare* listed perceptibly. Peter Soriano opened his door on the port side and watched as a narrow horizontal crack across the twenty-three-metre breadth of the hull, between the No. 4 and No. 5 holds, began to widen to form a V. The single propeller, meanwhile, continued to churn as the front half of the ship literally sheered off. The stern, ploughing an erratic course through the sea, came upon the 100-metre-long bow, now turned around and pointing 180 degrees to the rear. In the darkness and confusion, some of the frightened crew mistook the bow for another ship that had somehow arrived to rescue them. "We have a rescuer!" they shouted — until struck by the icy realization that it was a piece of their own broken vessel. As the bow disappeared behind them, the forward-moving stern began to sink, and the poorly trained crew made a mess of launching the two forty-man fibreglass lifeboats and two life rafts, twenty-man and ten-man models, all of them located aft. (One six-man life raft was located on the now-departed bow section, but none of the crew was in that section of the *Flare* when it broke in two.) The lifeboats had been secured to the *Flare* with extra lashings because of the poor weather, making it difficult for the crew to release them, and they turned turtle, landing deck down in the water. The life rafts inflated once tossed into the sea, but the securing rope on one chafed and the raft blew away. None of the crew wanted to jump immediately anyway for fear of being chewed up by the still-spinning propeller. There was no sign initially of the captain or first mate and therefore no order to abandon ship, so crew members simply waited, some crying, some praying. Chaos reigned. Eventually someone located the captain, who was asked whether the crew should abandon ship. "No, not just now," came the reply.

After about twenty minutes, the stern was virtually submerged, and the crew — with salt water at about one degree Celsius lapping at their toes — dove into the sea to make their way as best they could. "Chief, goodbye," Peter Soriano said to the chief engineer as he began to swim. No one wanted to wait until the stern actually sank for fear of being caught in the downward suction. The time from the loud crack to final abandonment was somewhere between twenty and forty minutes. Many of the crew wore only shorts, underwear and T-shirts, though many had put on one of the ship's thirty-five lifejackets, which were green or black and therefore difficult to see. Soriano had a thin shirt, pyjama bottoms and a windbreaker underneath his lifejacket. No one managed to squeeze into any of the ship's six immersion suits, which would have provided a much higher level of insulation — up to twelve hours until unconsciousness in waters this cold. (Six suits is the minimum number required under international regulations, even though the crew numbered more than four times that.) The survivors later said they didn't even know where the suits were stowed.

As the broken stern plunged 360 metres down to the ocean floor and its thick, heavy fuel gurgled lazily to the surface, the terrified crew battled waves as high as a two-storey building. Six of the strongest made it to one of the two overturned lifeboats and clung to its slimy back. The rest became pale points of luminescence as the rescue lights on their life-

The broken bow of the MV *Flare* showing a collapsed deck crane. TSB

jackets burned briefly. There was screaming for a few minutes, but the terrified cries soon subsided. The only sound remaining was the oily wash of sea and the icy breath of wind. Death was mercifully swift for most of the crew.

Waves smashing against the overturned lifeboat threatened to sweep off the six desperate survivors, now blackened by oil and cooling off rapidly. Two of them lost the battle after clinging for two or more hours. Soriano remembers seeing the bubbles of a last breath after a big wave knocked one man to his death. He tried to hold onto another survivor, the radio operator, when a wave knocked the man loose. But Soriano's left hand simply had no strength any more. (Unknown to the remaining survivors, this man became badly entangled in ropes underneath the boat.) The smell of the oil made all of them dizzy and disoriented, and some even drifted off to sleep for short periods. Remarlo Napa prayed for rescue, focusing his mental powers on seeing his new child, due to be born the following month. He grasped a rope as the cold sapped his rationality but left his strength intact. For five hours the men clung until, miraculously, they saw the piercing search beam of a helicopter slowly approaching, and they felt the hammering sound of its blades in their chests.

■ ■ ■

Marine and air search and rescue in Canada is organized around five operations centres that receive and act on distress calls, develop search plans and dispatch aircraft and ships. The three main centres are compact, drab bunkers located in Victoria, British Columbia, Trenton, Ontario, and Halifax, known as rescue co-ordination centres or RCCs. They are part of a world network of similar centres, created and operated under international agreements dating from the late 1940s. Canada's east coast also has two so-called sub-centres, in recognition of the intense marine activity that spreads from the St. Lawrence River through the Grand Banks and beyond. These sub-centres, located in Quebec City and St. John's, are smaller command posts staffed by coast guard radio specialists and marine rescue experts. Each of the three larger centres handles roughly similar caseloads, but the kinds of distress calls differ at each location. RCC Victoria is the only centre that must deal regularly with mountainous terrain, where hikers, private helicopters and small planes disappear regularly. RCC Trenton's bread-and-butter is pleasure-boaters on the Great Lakes during the summer months, when rescue activity soars.

RCC Halifax handles the bulk of the fishing-boat cases, as the Atlantic fishery remains the busiest in the country despite the high-profile collapse of cod and other stocks. The same swath of land and water also lies beneath a major passenger-jet corridor between Europe and eastern North America. About 650 international flights transit the Halifax search-and-rescue region every day in the peak summer season, and there are about 207,000 overseas flights each year. Fortunately, air disasters on the scale of Swissair Flight 111 are extremely rare. More frequent are major shipping disasters, often occurring in the brutal winter months when storms are at their worst but cargo traffic continues unabated. Once or twice every winter, a big ship gets in trouble, and RCC Halifax is suddenly galvanized into action.

The Halifax centre's area of responsibility, defined by international air and sea agreements, extends deep into the North Atlantic — almost to the Azores — and northward to cover most of Baffin Island, a total area of about 4.7 million square kilometres. About eighty per cent of this region is covered by water, and much of the North Atlantic area is well out of helicopter range. Every federally owned ship and aircraft can be pressed into service as a search-and-rescue asset, but RCC Halifax relies primarily on specially equipped Labrador helicopters stationed at rescue squadrons in Greenwood, Nova Scotia, and at Gander, Newfoundland; on Hercules transports with rescue gear at Greenwood; and on coast guard ships harboured at various points throughout Atlantic Canada. Civilian aircraft and ships, organized and trained as rescue auxiliaries, are also increasingly called upon. RCC Halifax coordinates more than 400 cases a month in the summer and as few as sixty a month in the winter. But the winter cases are easily the most taxing, often involving bulk carriers hammered by dangerous storms.

The Halifax bunker itself is an unassuming place deep in the bowels of a modern brick building near a top-secret, lead-lined navy operations room. A picture window looks out onto the navy dockyards, where frigates and destroyers are berthed, but the work stations are located well back from the daylight glare, and the crews prefer to keep the lights low the better to see their computer screens. The facilities are under the authority of the navy, but the operation itself is jointly run and staffed by the military and the coast guard. In the day, the centre is run by four people, each on a twelve-hour shift: an air controller and an assistant, both of them military officers, and a marine controller and a radio operator, both from the coast guard. The evening shift — where everyone trades their uniforms for blue jeans — drops the assistant air controller. The work stations integrate several sophisticated programs developed in Canada that plot search plans and calculate positions, and there's a central chart table with plastic overlays that help define search areas the old-fashioned way, with felt pens and rulers.

The room might as well be a stock trader's office or an insurance agency — except when a big rescue case erupts. Then the space is electrified, drawing in extra staff and TV cameras. Just outside the centre is a media room and during a big rescue a large television that hangs from the ceiling is flipped on as staff watch how their performance is being portrayed to a national audience. Reporters are generally poorly versed in search and rescue, and their stories are usually faithful regurgitations of the regular briefings provided by public affairs officers. That ignorance often serves the military and coast guard well.

Everyone who works at RCC Halifax has an extensive background in search and rescue, either aboard coast guard vessels or at the five search-and-rescue military squadrons across Canada — in Comox, British Columbia, Winnipeg, Trenton, Greenwood, and Gander. In addition, staff must take specialized training at the sprawling coast guard college in Westmount, Nova Scotia, a suburb of Sydney. This facility, resembling an oceanfront country club, was overbuilt in the heady days of federal government largesse to Cape Breton and found itself without enough students in the 1990s to justify its operation. It has since marketed itself internationally as a training centre for marine safety, environmental response and search and rescue, drawing students from Taiwan, Jamaica, Korea, Chile and other countries. In 1992, the college also took over responsibility from a coast guard facility in Cornwall, Ontario, for the training and testing of staff for Canada's three rescue co-ordination centres and two sub-centres. Twice a year, in thirteen-day courses, groups of eight students who have search-and-rescue backgrounds are put through the wringer.

Training includes learning how to answer the telephone and how to listen carefully — in a distress situation, a few moments' contact may be all that's available. Whoever answers the telephone must be able to quickly extract comprehensive information about a crisis from often-panicky callers. The heart of the course is time spent in one of two mockup bunkers, where four students prosecute a rescue case. These stripped-down rooms — pale yellow walls, sloping ceiling, picture window with turquoise curtains and a big wall-map of the East Coast — are scenes of intense emotion. Four work stations are all wired into a hidden control room, or "feeder" room, next door, from which details of a real-life case are piped into the bunkers. The feeder room has a library of cases to choose from, including some of the biggest rescues, though their names are scrambled. One case, for example, labelled *Ikamar* is based loosely on the 1994 sinking of the ore ship *Marika 7*.

Each such session might begin with a telephone call, and, as the cases develop, complexities are introduced. Often a second and third mock case is begun while students are still scrambling to resolve the initial problem. If things are going too smoothly, someone in the control room can throw in a monkey wrench — say, the grounding of a key search aircraft — that forces the team to react quickly. All of their activities are closely monitored, using video cameras that can zoom in on a sweating brow or a twitching eye. Hidden microphones record breakdowns in teamwork and help weed out misfits. Public affairs officers call in regularly to mimic nosy, obnoxious reporters, in order to teach students to deal with traps laid by the media. The experience can be so real and so complete that students sometimes panic, says Jean Maillette, superintendent of search and rescue. The experience is made even more intense when the feeder room decides to speed up a session, to make the team prosecute the five-day search for *Ikamar*, for example, in just two days.

This unusual facility is unique in the world and customized to the specialized needs of Canada's rescue co-ordination centres. Other countries employ a far more piecemeal approach to search and rescue. In Britain, for example, marine and air services are provided in separate HM Coastguard and Royal Air Force centres, more than twenty such facilities for a country a fraction the size of Canada, which has just five. The Australian system similarly allocates responsibility for air and marine rescue to separate agencies, each with its own agenda. American rescue co-ordination centres are operated primarily by the U.S. Coast Guard, which owns fleets of helicopters and fixed-wing aircraft in addition to its vessels. These centres are co-located with anti-drug and immigration operations, so that search and rescue becomes just one of several priorities. Canada, by contrast, operates highly integrated rescue co-ordination centres, with relatively seamless co-operation between the coast guard and the military. Their one and only mission is search and rescue, along with humanitarian relief such as medical evacuation. The result is a tightly focused, efficient use of taxpayer dollars to patrol enormous stretches of territory.

RCC Halifax itself was established in 1947 initially as a Royal Canadian Air Force operation dedicated primarily to helping stricken military aircraft. Its role gradually broadened over the years, and by 1960, coast guard personnel were added to the staff to better co-ordinate responses to marine incidents. In 1964, Canadians were outraged to learn that a tug from the Soviet Union had to tow a disabled Canadian fishing vessel off Liverpool, Nova Scotia, because the Canadian Coast Guard was not allowed to respond to such incidents under federal policy. The rules were soon rewritten, and the number of marine cases soared, forcing coast guard staff increases at all centres. Since then, the centres have been models of inter-departmental co-operation, operating twenty-four hours a day, year round. One mayday — night or day, from land, sea or air — immediately triggers a broad range of resources, from ships, planes and helicopters to radio specialists, medical personnel and computer experts. But first, that mayday has to be heard.

■ ■ ■

Ann-Margret Basha, alone in a dark office building in the middle of the overnight shift, began to question her decision to quit teaching high school for this so-called career in the coast guard, which she had jumped to six years earlier. "There's got to be more to life than this," she thought as she attended to weather and radio routine. Basha was almost ten hours into her lonely twelve-hour stint in the radio operation room of the

Stephenville coast guard facility, on the second floor in the back. The building, on a snow-blown government wharf in western Newfoundland, was completely deserted. Basha made brief radio contact with her counterpart at the Sydney, Nova Scotia, coast guard station, complaining that it was turning out to be one of the longest nights of her life. She began pacing around the room, trying to kill the last two hours, when suddenly a radio speaker squawked: "Somebody help us, somebody help us, mayday, mayday, mayday." Electrified, Basha jumped to the console and saw from a display light that she wasn't receiving the call directly. Instead, the transmission on Channel 16, a standard VHF marine channel, had been picked up at an unmanned receiving station on the island of Ramea about 150 kilometres to the southeast. The call had been received, then instantly retransmitted by commercial microwave to Stephenville. The Ramea site, just off Newfoundland's south coast, can pick up transmissions from a minimum radius of seventy-five kilometres. But at night, when atmospheric conditions are better and general radio chatter is quieted, the tower at Ramea can pick up signals from twice that distance and more.

"There was no doubt as to the urgency," Basha, then thirty, recalled about the distress message, which came in at one minute before five local time. "It was a definite — it wasn't a prank call." She waited three or four seconds to see whether any coast guard station that might be closer would answer, standard procedure to ensure she did not walk all over someone else's transmission. None answered, so she responded: "What station calling mayday? This is Stephenville coast guard radio on Channel 16. Say again, over." Then followed a short burst of two voices attempting to relay a position. "Thirty-seven," said a panicky man shouting in the background, then another voice saying "46 37 north (garble) 54. 36 37 15 north." The first voice then broke in with "No. Latitude." Basha, jotting notes furiously, asked for confirmation of the position and the number of persons on board, but there was no response. No vessel name, nothing about the nature of the emergency, no word about abandoning ship, nothing about the crew members. The entire exchange was over in about twenty seconds. "It came in quite clear," she said later. "There's not too much activity on VHF at that hour of the morning, so it was a fairly strong signal, but broken. It was loud. It wasn't consistent." The voices on the mayday message had an accent of some sort, perhaps Newfoundlanders on a longliner that had gotten into trouble. When it became obvious there would be no more transmissions from the mystery ship, Basha called the coast guard stations at Placentia, Newfoundland, and

Sydney to determine whether they, too, had heard the mayday. "Did you copy any of that?" she asked, but each time the answer was no.

Basha immediately issued a "mayday relay" on Channel 16 and on a longer-range marine channel, alerting all ships in the area to the emergency, giving them the approximate location and asking them to render all assistance. This message, repeated three times, was as much as she could offer without even the name of the vessel. (The long-range relay was picked up as far as Norfolk, Virginia, 2,100 kilometres to the southeast.) Basha then called Merv Wiseman, who was staffing the overnight shift at the Marine Search and Rescue Sub-centre in St. John's, located in the same room as the coast guard's radio operations centre, a brick building at the southwest corner of the harbour. Basha gave Wiseman the position as best she could make out from the sputtering transmission, 46 57 north latitude, 56 51 west longitude. Wiseman alerted the rescue co-ordination centre in Halifax and set about trying to locate ships in the area that might be able to assist. Meanwhile, Basha called her boss at home, and after about twenty-five rings, Derek White groggily answered, snapping to at the word "mayday." Arriving at the office within thirty minutes, White went to another section of the building where tapes of all transmissions are kept. Basha had not been able to leave her post to play back the tapes and so relied solely on her initial interpretation of those few seconds of crackling radio contact. White first had to record the master tape onto a smaller and more versatile machine, which could slow down the transmissions. He replayed the dubbed tape several dozen times, finally revising the latitude to 46 37.15 north, and the longitude to nothing more precise than perhaps 54 west. The remote Ramea site unfortunately was not fitted with specialized direction-finding equipment that might have helped pin down the position. And because of the inaccessibility of the tapes, fifty-five minutes had been lost trying to determine the location.

Luckily, an American satellite receiving station had detected a separate distress alert about three minutes after Ramea signal. The alert was a burst of recorded information from a high-tech bridge radio that finally provided the name of the ship, the MV *Flare*. The brief transmission also gave a position reasonably close to that distilled from the garbled mayday call — but with one problem. The satellite call indicated the *Flare* had been in that position about twenty-four hours earlier, suggesting it was now about eighty kilometres south of the mayday position. The current location was therefore uncertain, but at least they had the name of the ship. The U.S. Coast Guard rescue co-ordination centre at Norfolk, Virginia, relayed this new satellite information to Halifax nine minutes after the Ramea call. In St. John's, meanwhile, Wiseman had contacted what appeared to be the only ship in the area, the MV *Stolt Aspiration*, a Liberian-registered chemical tanker, which was at least thirty-three kilometres away from the distress position Stephenville had estimated. The ship had also heard the brief mayday on Channel 16. With westerly winds gusting up to seventy-five kilometres an hour, the captain of the *Stolt Aspiration* estimated it would take about six hours to get to the *Flare*. The closest navy vessel, HMCS *Montreal*, was off Halifax about 600 kilometres to the east.

The emergency clearly needed a much faster response. As officials checked the Lloyd's online catalogue of ships to get a detailed description of the *Flare*, RCC Halifax ordered a Labrador rescue helicopter from its base in Gander to fly to the mayday position that Basha and White had worked out. Here, finally, was a major sea rescue involving a bulk carrier that was close enough to the Labrador bases to allow them to participate. Cruising at an altitude of 150 metres, the rescue chopper was asked to fly what is known as an expanding square — a kind of outwardly moving spiral, except the spiral is a square using lines of latitude and longitude for ease of navigation. A Hercules aircraft from Greenwood was also ordered to check the alternate position given in the satellite distress call. The Herc was then to proceed north to the mayday position, then to head due east. A fisheries patrol aircraft, already scheduled for a pollution-watch flight from St. John's, was retasked. The green twin-prop King Air, on long-term charter from Provincial Airlines, would instead look for the *Flare* by traveling west from Cape Race, the southeasternmost tip of Newfoundland, along the same latitude as the Hercules. Hopefully, as the Hercules and King Air headed toward each other, one of them would spot the *Flare* before they met. The King Air's sophisticated radar was a particularly powerful tool, normally used to spot and record fishing vessels as it watched for illegal activity. On the sea, the coast guard dispatched three of its own ships and two cutters and called in a French navy patrol vessel from St-Pierre-Miquelon. Rescue officials also asked naval intelligence in Halifax whether they had a fix on the *Flare*, and, using satellites and other means, the top-secret centre provided a location near the purported mayday position. The Halifax rescue centre also asked the navy frigate *Montreal* to join the search, even though the vessel was off Halifax carrying out sea trials and at least ten hours away from the scene. Two other commercial ships farther away than the *Stolt Aspiration* were also ordered to help, as required under international law.

The trouble was, no one knew exactly what they were looking for. The *Flare* was described as a 181-metre cargo ship with a black hull, red deck and white superstructure at the stern. But had it sunk? Try as they might, Canadian and international radio operators could not reach the ship via either satellite or standard VHF frequencies. "We have received a . . . distress alert from your vessel. Please advise your current position, course, speed and the nature of your problem," RCC Halifax asked the ship's captain, listed as Georgoulis Zannis. There was no response. Rescue officials contacted the ship's agent in Montreal and were told there were twenty-six persons on board and that the last contact with the ship had been three days earlier, on January 13. Soon after, the company manager in Athens called to say *Flare* was equipped with two orange forty-man lifeboats and three life rafts, along with six survival suits. Here at least was something to pass on to the search aircraft and ships, which were now being joined by two more Labrador helicopters and another Hercules as the potential severity of the distress became clearer. Watch for two lifeboats and three orange life rafts, they were told, and possibly immersion suits.

But where was the rescue beacon? All large commercial ships are required by international law to carry an emergency position indicating radio beacon, or EPIRB. This is a transmitting device about the size of a large thermos that normally sits outside the bridge. If a ship sinks, the battery-powered device is meant to float free and begin emitting a signal on a special frequency, 406 megahertz, reserved for emergencies. The signal — emitted in a regular burst rather than continuously — carries a unique code identifying the EPIRB's owner and is intended to be received by a group of weather satellites in low polar orbits. The satellite system, known as SARSAT for "search and rescue satellite aided tracking," is operated jointly by the United States, Canada, France and Russia. Begun in 1979 and declared operational in 1985, the system is a model for international co-operation, much like the Antarctic Treaty. Canada, one of the founding partners, provided the electronic packages for the satellites and maintains ground stations in Edmonton, Churchill, Manitoba, and Goose Bay, Labrador. A mission control centre at Canadian Forces Base Trenton, Ontario, processes the signals and alerts the appropriate rescue co-ordination centre.

When an EPIRB starts to transmit, the signal is received by one of the weather satellites, which employs the Doppler effect to pinpoint location. A familiar example of the Doppler phenomenon is the changing pitch of a locomotive horn as it approaches the listener and as it then moves away. Sound waves become more compressed as the train moves closer, then they decompress as it moves away. A similar effect occurs when a satellite approaches and moves away from a radio transmission source. By calculating how much the frequency is compressed and decompressed, a location can be calculated — often to within metres. And because each beacon is registered, rescue officials can quickly check with the owner to see whether any signal is a false alarm. There are more than 100,000 such beacons on ships worldwide, about 2,000 of them in Canada. They represent a simple, effective alerting system that has saved thousands of lives in the last fifteen years. Worth about $2,400, the EPIRB is a low-cost lifesaving device that can quickly alert officials to the precise location of a sinking and get a rescue aircraft on the scene without delay. And once on scene, an aircraft or ship can then home in on the EPIRB's second signal, continuously broadcast on the 121.5 megahertz frequency, to speed the search process.

The *Flare* was equipped with a modern EPIRB, as required by all ships in its class. The device, built by Ceis Espace, had recently passed inspection, and its battery was deemed good until March 1999, more than a year away. A CBC television documentary on the disaster, *Tragedy at Sea*, broadcast on February 2, 1999, suggested that the EPIRB had fallen from its mounting bracket on the radar mast during rough weather and broken, and it had not been repaired. This at least was the view of one of the rescued sailors interviewed. "The case of the EPIRB is rusty already," said Cyrus Ferraren. "So the EPIRB was fallen down and broken and not replaced." But these devices are never located in a place as inaccessible as the radar mast because they sometimes need to be triggered manually in an emergency. A more typical location is on the bridge deck — and information the *Flare*'s owner faxed to the St. John's rescue sub-centre showed that that was precisely where the EPIRB was mounted, on the starboard wing. Yet throughout the rescue and after, the device was silent. None of the survivors reported seeing it in the water. A victims' lawsuit later claimed it had been improperly stored inside the bridge, though there was no proof. "This has been one of the aspects that has bothered me from day one," Bill McOnie, chief investigator in the *Flare* case for the Transportation Safety Board of Canada, said later about the mysterious EPIRB failure. "Why it didn't activate, no one knows. We never will." The *Flare* search, it seemed, would have to rely entirely on the garbled mayday call and the confusing radio-satellite transmission.

Soon after arriving at the general search area, one of the Greenwood Hercules aircraft dropped two high-tech missiles into the rolling seas.

These Canadian-developed devices — known as self-locating datum marker buoys, or SL-DMBs — were each tucked inside a metal tube. As the tubes hit the water, the impact triggered a carbon-dioxide cartridge that inflated a triangular orange air bag attached to a cylinder inside the tube, pulling the cylinder free and allowing it to float at the surface. The first cylinder that was released was designed to mimic the drift of a life raft caught in the wind and little affected by the current. The second cylinder released had a drogue chute attached that slowed its movement in the water, mimicking the drift of a person caught in the current and little affected by the wind. Radio signals from these devices would then constantly broadcast their current positions, via satellite, using the Global Positioning System. Rescue officials could thus make highly educated guesses about how the search area was expanding over time.

The strange-looking devices, dubbed "dinosaur condoms" by some of the crews, had been tested in the *Vanessa* search-and-rescue mission less than three months earlier. Canadian Coast Guard-developed computer software, the Canadian Search and Rescue Program (CANSARP) had been used to predict the location of life rafts and persons in the water — but it was only as reliable as recent weather information and historical data on ocean currents. The program, under development since the late 1980s and available at all Canada's rescue co-ordination centres by 1991, drew on pioneering work by Bernard Koopman, the American who developed techniques for predicting the location of enemy submarines and whose work was adapted for ground searches. Koopman's complex formulas were ripe for computerization, and the Canadian Coast Guard hired a Vancouver company to develop the software. This valuable tool can help predict the drift of twenty-six different types of objects in the water — powerboats, sailboats, life rafts, people, etc. — using current and wind information. Weather data in the form of a forty-eight-hour forecast for the various regions is obtained from the Canadian Meteorological Centre in Dorval, Quebec, every six hours. Information about ocean currents and tidal action is obtained from historical databases, and it is constantly being expanded with real-time information from semi-permanent sensors in the oceans. Based on this data, CANSARP outlines a search area, which can then be overlaid with a search pattern appropriate to an aircraft or ship, such as the expanding square pattern used in the *Flare* operation. This diagram can then be faxed directly to search crews to give them a clear picture of the task at hand.

CANSARP quickly became an enormous timesaver as it went through various refinements in the 1990s, but unfortunately, poor quality weather input gives poor quality predictions; its effectiveness is directly related to the accuracy and timeliness of the data it draws upon. In December 1994, the program was singularly ineffective in predicting the location of the survivor of the sinking of the *Salvador Allende* about 900 kilometres south of St. John's. Alexander Taranov was found about 112 kilometres southeast of the prediction, mostly because there is only sketchy information about currents and winds that far out in the North Atlantic. The new high-tech datum marker buoys developed by the military and coast guard are one answer to this vexing problem because they provide a steady stream of fresh information about how an actual object is drifting in the search area. This important data proved vital in the *Vanessa* rescue: CANSARP by itself made a poor prediction of the location of a person in the water, but the juice-can-sized SL-DMB dropped into the water helped RCC Halifax adjust its search strategy, resulting in the rescue of one man who had been adrift in the sea for nineteen hours. In the *Flare* rescue, too, CANSARP could draw on fresh SL-DMB data to keep it accurate. This was only the second major Canadian mission using this new technology. The survivors of *Flare* — if there were any — would be among the first beneficiaries in the world of a pioneering Canadian technology.

The federally chartered King Air, meanwhile, had taken off from St. John's airport, travelling almost due south along the eastern coast of the Avalon Peninsula. At Cape Race, the twin-prop aircraft turned due west, hugging the line of latitude given in the *Flare*'s mayday call. Shortly after passing to the south of St-Pierre-Miquelon, the King Air's radar picked up an unidentified target a little off the line of latitude. The crew duly reported the sighting to the Hercules crew that had been travelling the same line of latitude coming from the west. The King Air then headed north to carry out an expanding-square search to the west of St-Pierre, assuming the radar sighting it had provided would be investigated. The Hercules was now acting as the on-scene-commander, closely co-ordinating the search by communicating with all the aircraft in the search area, which now included an Aurora patrol plane and two Labrador helicopters, as well as with various ships. The Hercules radar picked up the same unidentified target as the King Air, but the busy crew apparently assumed it was another rescue vessel helping in the search and did not fly over it to investigate.

Two hours later, with nothing to show for their sweeps of the area to the north, the King Air crew returned to their starting point south of St-Pierre and asked what had been done to identify the mysterious radar target. The Hercules crew claimed not to have received the initial message from the King Air and now told the King Air crew to investigate. A few minutes later, the small aircraft was looping around the bow of the *Flare*, getting the first look at what had happened in this bizarre marine disaster. The hull had clearly cracked in two, and this front half remained afloat more than five hours later, drifting free. The King Air circled the bow seven times, then carried on east, turned west again past the bow and soon spotted the giant fuel slick. Their pollution-watch flight had turned up an actual oil spill after all, though under circumstances no one could have imagined. There at the western end of the shimmering, rainbow-hued patch was the overturned lifeboat with four people clinging to it. Only a few minutes later the Labrador crew happened independently upon the same scene and carried out their textbook rescue.

The slip-up in checking out the radar target left the four survivors in the frigid ocean an excruciating two hours and thirteen minutes longer than was necessary. The incident only came to light when the King Air crew was debriefed after the mission at the rescue centre in St. John's. Unfortunately, the information came too late for post-mission investigators to check the in-flight voice recorder on the Hercules, which had by then been erased. The military did determine, however, that there was no mention of the King Air information in the Hercules flight logs. An investigator with the Transportation Safety Board of Canada, John Mein, later called the error an "oversight" and attributed it in part to a busy cockpit trying to digest a lot of information from various aircraft and ships. "Being on-scene commander, it's a very hectic job for them," Mein said in an interview. "Because they've got lots of communications coming through the radio and whatnot, on top of doing their own job in the plane and trying to keep a lookout. . . . It was an oversight on their part." A separate joint military-coast guard investigation called the episode an "anomaly" and insisted that the King Air message had never been received by the Hercules, without explaining why the Hercules's own radar sighting was mistakenly thought to be a rescue vessel. Mein nevertheless would conclude that the rescue delay did not contribute to the deaths of the two crew members who had slipped off the lifeboat some hours before the Labrador showed up. Ironically, the Hercules aircraft itself got into a jam when it ran low on fuel; it had to bypass a landing at the airport in Stephenville and make a low-fuel emergency landing at Deer Lake, Newfoundland.

The discovery of the survivors' lifeboat gave the CANSARP operator at RCC Halifax an important additional piece of information, an actual drift object with a known location about six hours after the disaster. This data was fed into the program, and then it was run in reverse — or backdrifted — to determine the precise point of the sinking. From that calculated point, the program could then predict how far and where a life raft or a person in the water would have drifted. This information was passed on to search planners, who recalibrated the tasks assigned to the ships and planes operating in the so-called "hot area," the prime search area. Another self-locating datum marker buoy was dropped at the location of the overturned lifeboat to provide more data about how things were drifting apart. But, like many things technical, the program inexplicably broke down at a critical moment. Wind information that was supposed to be updated every six hours from Dorval, Quebec, was not being read: CANSARP was stuck with readings that were at least twenty-one hours old. A staff member of the Canadian Coast Guard College near Sydney was ordered to stay as late as it took to resolve the problem.

Eight hours into the rescue, the question naturally arose at RCC Halifax about how many of the *Flare*'s crew could have survived in water and air temperatures hovering around the freezing point. The four men pulled from the lifeboat were barely alive after about six hours — and they were in the frigid air, where the body stands a better chance against hypothermia because water is a much more efficient conductor of heat away from the body. (Sweat, for example, is the fastest way to rid the body of excess heat.) But just how do you reliably predict human survivability in the ocean?

The problem, in fact, had been analyzed in detail in some path-breaking military research at the Defence and Civil Institute for Environmental Medicine in Toronto. One of the lead investigators was a physicist, Peter Tikuisis, who had painstakingly developed models of the human body struggling against the outflow of heat. His study was initially undertaken for the Canadian air force, which wanted to replace its helicopter-equipped rescue units on the Prairies with ground-based vehicles. In order to know where to station the vehicle crews, military planners needed to know how long downed airmen could survive in the cold before succumbing to hypothermia. An American scientist, Eugene Whistler, had already developed theoretical models for how the human body transfers heat in a thermally neutral environment. But no one had ever developed a model for the ability of humans to survive in extreme temperatures, which trigger the special physiological responses of shivering and unconsciousness. In 1994, Tikuisis and colleague John Frim published a paper outlining a model for exposure to cold air. The premise was that the human body in a resting state naturally produces heat that is gradually dissipated depending on the thickness of the fat layer under the skin, on the amount and type of clothing, and on the temperature of the air. Shivering, the body's answer to potentially life-threatening cold, creates heat and can delay the inevitable lowering of the body core temperature somewhat, depending on the body's store of energy.

Tikuisis and Frim drew partly on a Canadian experiment in which military volunteers sat naked for three hours at five degrees Celsius in an artificial wind. The conditions induced mild hypothermia, but, because medical ethics preclude endangering life, the physiology of extreme hypothermia had to be extrapolated from the data rather than observed. Other quantifiable factors were incorporated into the model. A chubby person with, say, twenty-nine per cent body fat stands a far greater chance of survival — about double — because of this natural insulation than a lean person with, say, eleven per cent body fat. Injuries involving blood loss or trauma can inhibit the body's ability to fight against the cold. Tikuisis was also given access to the medical files of the men and women whose Hercules transport crashed near the northern military outpost of Alert in 1991. Most of the group survived thirty-six hours in temperatures averaging minus twenty degrees Celsius, and their experiences appeared to corroborate the model's predictions.

Tikuisis expanded the research — primarily out of personal interest — to survival in cold water. His broader intention now was to give search-and-rescue planners another factor to consider when deciding whether to call off a mission when survivors have not been located. Most such missions typically continue far longer than any realistic expectation of survival, partly for humanitarian reasons such as assuring relatives that everything possible has been tried. But planners also had few objective guidelines about when to call off a major search, which can itself be risky to rescue crews. A medical formula indicating the limits of human endurance could provide clearer guidance about when to stop searching.

As with the cold-air model, Tikuisis was able to draw upon experimental results using military volunteers. Thirteen men were asked to exercise for five hours without food and then endure exposure to the cold for four and a half hours, wearing T-shirts and shorts. For the first half hour they sat in a room with the air at ten degrees Celsius, then for the next four hours they were sprayed on their backs with a stream of water at the same ten-degree temperature. Rectal thermometers measured their body-core temperatures. Only three men endured the full assault.

Cramping, pain and body-core temperatures dropping below thirty-five degrees Celsius eliminated the remainder. A company in Dartmouth, Nova Scotia, was contracted to provide insulating values for various types of clothing, which they calculated by dressing up a mannequin embedded with thermal sensors. In 1996, Tikuisis assembled all of these factors into a mathematical model that he then converted into a Windows-based program for an ordinary personal computer. This software was distributed to every rescue co-ordination centre in Canada. Plug in the values — height, weight, body fat, age, sex, water temperature, clothing, etc. — and the program will compute the number of hours until the body core hits that thirty-degree Celsius threshold at which a person normally tips into unconsciousness.

Users of the program are encouraged to produce optimistic values, that is, they should assume the victim is fresh and uninjured. The program also factors in gender differences, because females tend to have a higher body-fat content and therefore enjoy a theoretically greater prospect of survival though they tend to be shorter and lighter, which are negatives for cold-water survival. But the program cannot measure the will to survive which, for example, kept Remarlo Napa alive and conscious at a core temperature of 26.8 degrees Celsius. Tikuisis's model, which is unique in the world, has been used by the U.S. Coast Guard and is being upgraded to take into account injuries to arms and legs caused by cold to help rescue planners determine whether victims will be able to help themselves.

Officials at RCC Halifax entered some typically optimistic values into this model for potential *Flare* survivors: a heavy, twenty-five-year-old male of medium height, fully immersed in the water, wearing a heavy parka, with seas at two degrees Celsius. The prediction: 4.1 hours to unconsciousness. The four survivors at the St-Pierre hospital had indicated that the ship went down quickly, soon after the mayday call, so anyone in the water would have been there for eight hours. No one had made it into the life rafts, which had all been accounted for. And no one had managed to squeeze into one of the six survival suits, which Peter Tikuisis's program said would keep a twenty-five-year-old person alive for twelve hours. The likelihood of survival was thus virtually nil.

Although hope is never entirely abandoned, this search-and-rescue operation was now quietly transformed into a body-recovery mission. The media and relatives would be offered hopeful words, but it had become clear that the four men plucked from the overturned lifeboat would be the only survivors from the crew, now determined to number twenty-five rather than the twenty-six the Montreal agent had initially indicated. Sure enough, soon after the sea-survival prediction, the coast guard cutter *W.G. George* reported that it had recovered two "blacks" — the standard code word for dead bodies. The cutter brought aboard two more bodies soon after, but all of the deceased were so contaminated with oil from floating in the slick that recovery ceased and the boat simply remained alongside three more bodies after marking the spot with an electronic buoy. Two of the Labrador helicopters hoisted up eight bodies but also ceased the operation because of heavy oil contamination. They both delivered their dead to St-Pierre. These helicopters and coast guard ships would later need extensive oil decontamination.

Montreal, finally on scene after its long voyage from Halifax, was appointed the unofficial morgue; it recovered the three bodies bobbing alongside the *W.G. George* and transferred on board the cutter's four other bodies. One of the dead men was carrying a crew list, confirming the total number aboard at twenty-five. All of these seven bodies wore only light clothing, such as cotton pants. Two did not even have shoes on. Two wore leather jackets, one had a jean jacket, and two wore thermal underwear. Their lifejackets were old-fashioned, with little buoyancy; a later laboratory test of one such lifejacket showed it did not meet international standards, though the *Flare* was allowed to carry them because it was older than the regulations. Oil covered the flashing rescue light on one of the jackets, making it impossible to see. The corpses were stripped, cleaned of their coating of thick oil and placed in the ship's refrigerated dairy room below decks. The remains and belongings were all meticulously tagged. The locations of the bodies were also entered into the CANSARP program as actual drift objects to help verify and fine-tune the predictions, although there was in fact no prospect of finding anyone else alive.

An Aurora and a Hercules aircraft carried out the perfunctory overnight air search, while *Montreal* and another coast guard ship continued the sea search, relieving three commercial vessels that were thanked for their help and told to carry on with their intended journeys. Sea temperatures remained at two degrees Celsius, but the air temperature had dropped five degrees to minus seven, with winds at about sixty-five kilometres an hour. No more bodies were found, and the search was officially called off at noon the next day, January 17: four rescued, fifteen dead, and six missing and presumed dead. The case had stretched across almost 32 hours from the time of the first mayday call, eight times longer than any person could reasonably be expected to survive in water so cold. More than 20,000 square kilometres had been covered in the operation, an area three times the size of Prince Edward Island. Eleven aircraft had together spent fifty-six hours in the air. Five coast guard vessels, three commercial ships and the French navy vessel *Fulmar* had together spent ninety-nine hours searching the seas; *Montreal* burned about $83,000 worth of fuel in the search.

Montreal headed for Fortune, Newfoundland, where it was to meet the coast guard cutter *W. Jackman* and receive the eight bodies it was transferring from St-Pierre, but bad weather prevented the ships from accomplishing their task. Jurisdictional problems prevented *Montreal* from returning to Halifax with the bodies, since Newfoundland's senior medical examiner, Dr. Simon Avis, had to inspect them. Eventually *Montreal* delivered its seven bodies to the RCMP at Argentia, Newfoundland. The cutter *W. Jackman* left its own sad cargo with the Mounties at Fortune. All the bodies were later taken to St. John's for official determination of the cause of death. Two days after their rescue, the four survivors flew from St-Pierre to St. John's to identify the bodies of their dead crewmates. The names of the six unrecovered men were turned over to the RCMP as missing-person cases; their bodies still had not been recovered three years after the disaster.

Montreal headed back to its home port of Halifax, passing again near the location of the tragedy. The crew laid a wreath on the water in memory of the dead and observed a one-minute silence. In a telexed message to RCC Halifax, the captain cited Chapter 2 of the Book of Jonah, where Jonah prays to God from inside the belly of a great fish: "You have cast me into the abyss, into the heart of the sea, and the flood surrounded me, all your waves, your billows washed over me. . . . But you lifted my life from the pit, Yahweh, my God."

The sunken stern of the *Flare*, found on January 20 by a coast guard ship, continued to disgorge foul black oil. Search-and-rescue aircraft attempted to ignite the persistent slick to prevent it from contaminating sea life, but the fuel simply would not light. The oil flow would continue for another six months. The coast guard recovered the fibreglass boat that had saved the lives of Markovic and his crewmates; the companion boat washed ashore six months later at Lamaline, Newfoundland, its underside thickly crusted with mussels. The listing bow of *Flare*, meanwhile, continued its bizarre voyage southward, prompting the coast guard to issue a notice to all shipping in the area about this potential hazard. Almost four days after the disaster, the bow finally sank in about 100

The upside-down bow of the MV *Flare*, filmed by the remote-controlled submersible *Sea Rover*. TSB

metres of water near the western edge of the Banquereau Bank, off Nova Scotia. A salvage tug that had been shadowing the rogue vessel reported that it upended and sank vertically with the bow end up. Air could be heard hissing out of the cargo holds as the hatches popped. Six months later, a remote-controlled submersible — the 300-kilogram *Sea Rover* — was lowered on a tether to the site from the Canadian Coast Guard ship *Earl Grey*. The bow had flipped upside down, with the starboard side deck embedded in silt on the sea floor. The initial impact crater remained visible about thirty metres away from where the wreck had finally come to rest. The *Sea Rover*'s still camera malfunctioned, but its video cameras, guided by a mini-sonar, provided footage including an eerie sequence showing the vessel's inverted name. The videotape was sent to labs in Ottawa operated by the Transportation Safety Board of Canada, which was given the job of finding out what had gone wrong.

Early indications were that the ship's steel hull split in two because of "brittle fractures," a weakness in the metal typically occurring in older bulk carriers that have been pounded by stormy seas and by the hard loading of cargo over many years. The chief investigator, Bill McOnie, also said that the *Stolt Aspiration* — the cargo ship nearest the *Flare* that rushed to the rescue — reported seeing a giant rogue wave. These mammoth waves, up to twenty metres in height, readily form in the area between Cape North, Nova Scotia, and Cape Ray, Newfoundland. The *Flare* might well have been smacked by such a wave, causing the already weakened hull to crack open. The investigation quickly ruled out a design flaw in the ship. But the underwater examination of the hull, it was hoped, would help engineers know where to look for early signs of brittle fractures in similar ships. The search-and-rescue operation became the subject of dramatic television newscasts and newspaper reports at the time of the incident. But in the weeks and months following the tragedy, media attention naturally shifted to whether the *Flare* was too old to be plying the oceans and whether other aged ships should be ordered scrapped.

The board's final report, released more than two years after the disaster, confirmed the brittle-fracture hypothesis by drawing on photos of both the bow and stern. Small clamshell-shaped fissures that dotted the cracked ends of the vessel were an unmistakable sign of this catastrophic process. But the investigation also found that the ship was improperly ballasted. Aerial views of the floating broken bow showed an accumulation of ice and snow on the deck, clear evidence that the deck metal was colder than it should have been because the ballast tanks below were empty of water. Port records in Rotterdam also showed that the ship lacked the ballast recommended by the loading manual. Investigators later estimated that the ship carried only about three-quarters of its required ballast, allowing the ship to ride too high. The captain, on his first command of a bulk carrier, had neglected to ballast the ship properly before departing Rotterdam, perhaps in a misguided effort to save fuel on the transatlantic voyage. His decision had subjected the ship to excessive pounding in the North Atlantic, increasing the chances of a brittle fracture ripping the aged ship apart. Most of the crew, poorly supplied with life-saving equipment and rescue training, paid the ultimate price.

A post-operation report by the St. John's search-and-rescue sub-centre made a handful of recommendations, including the installation of direction-finding technology at coast guard receiving stations to help narrow the location of an emergency in the event of a garbled message. To prevent a repetition of the delayed play-back of the recorded twenty-second communication between the Stephenville station and *Flare*, instant replay facilities were soon installed. The Transportation Safety Board also expressed concern about the lack of search-and-rescue boats along the south coast of Newfoundland, the result of cuts to dedicated rescue resources in the early 1990s. The board's report recommended

Tell-tale patches of ice on the left deck of the MV *Flare*. TSB

that enough immersion suits be carried aboard ships for every crew member where hypothermia was a risk, and that better training in the use of life-saving equipment be provided. In addition, it asked that Canada promote awareness of the importance of proper ballasting and the dangers of hull stresses — which Transport Canada agreed to do in international forums.

The *Flare* operation had demonstrated how the Canadian search-and-rescue system can quickly gear up to meet the demands of a major marine disaster, from a lonely radio watch that triggered the mission to Canadian-made locator buoys, search software and survivability predictions; from the integrated rescue co-ordination centre that organized the operation to the Labradors and Hercules specially configured for search-and-rescue in the Canadian environment; from the coast guard captains manoeuvring in stormy waters to the highly trained helicopter crew faultlessly plucking half-frozen men from disaster. Sartech Tony Isaacs may have dangled alone at the end of his steel cable, but that taut line connected him to hundreds of professionals — and to a rescue culture — that dealt decisively and competently with the bleak environment of an ocean disaster.

The four surviving crew members and some relatives of the dead later sued the *Flare*'s owner, ABTA Shipping Co. Ltd., and the operator,

Trade Fortune Inc. SA, for a total of almost $9 million. The action, filed in Federal Court in Montreal, claimed that the two companies showed "wanton, wilful and contemptible disregard for the condition of *Flare* and the rights and safety of her crew." The companies did not adequately inspect, maintain and repair the *Flare*, it said, nor did they maintain the life-saving equipment or train the crew in its use. The survivors still suffered medical problems as well as post-traumatic stress syndrome, their lawyers argued. Remarlo Napa, who returned to the Philippines, told a reporter he wanted to be a farmer and will never return to sea. Peter Soriano, asked whether he'll ever work on a ship again, said, "Like an old one [ship] again? No. I'll never go like that. Not even a new one." Petar Markovic returned to the port city of Bar, Yugoslavia, where according to his lawyer he still suffers from medical problems. "There's no question he's suffering badly from a number of ailments, including post-traumatic stress disorder," said Rick Sproule of Montreal. Lawyers for ABTA Shipping Co. initially argued that the case should be transferred to Cyprus, the company's putative place of business. The crew had also signed an employment contract agreeing to have any disputes resolved in Cyprus, where, coincidentally, liability limits were a tenth of the Canadian maximums. The judge tossed out those arguments, noting that most of the documents ABTA had filed in the case originated from London, South Korea, Cuba and Greece rather than from Cyprus.

The fact that the captain had failed to properly ballast the ship for the transatlantic crossing appeared to bolster the case of surviving crew members. But Daniele Dion, the Montreal lawyer representing the owners and operators, said that ballasting was always the sole responsibility of the captain and that the companies could not be held liable. The company had spent hundreds of thousands of dollars on maintenance and spare parts in the year previous to the sinkings, Dion said, and the ship was in much better shape than the plaintiffs acknowledged. ABTA Shipping, which had bought *Flare* in 1987 from a Liberian company for US $1.7 million, stood to collect US $5.1 million in insurance, though lawyers for the victims successfully argued that most of the money should be held in Britain pending the outcome of various legal actions. In mid-2001, the lawsuits remained unresolved.

Many weeks after the rescue, the search-and-rescue squadron at Greenwood, Nova Scotia, received a hand-printed letter in the mail from Markovic, a relatively unusual event. Most of the people the squadron members save are profoundly grateful at the time but typically disappear into their own lives far, far away. Not so Petar Markovic, who had come to know his rescuers well while they visited him over two days in the St-Pierre hospital. "This letter . . . is one of the most difficult letters I've ever written. I simply can't find the words to express my gratitude for everything you did for me," he wrote. "When I speak about you I usually say that even the best movie director can't show on the screen the ability, courage, skill and equipment that you possess. I am perfectly aware of the fact that you saved my life twice. The first time, rescuing me from the freezing water, and the second time, in [the] helicopter, giving me the first aid. . . . I know that we were saved only because of your team work and cooperation with other participants in the action, so I would like you to send my gratitude to all of them." Markovic's letter has been framed and proudly hung in a hall at 413 Squadron's Greenwood headquarters.

The *Flare* rescue demonstrated the dramatic life-saving ability of helicopters. Finally, a major east coast ship disaster had occurred within the rescue range and weather envelope of a Canadian Labrador, allowing its highly trained crew to show their tenacity and skill in a difficult operation. The fading survivors clinging to their lifeboat were near death and utterly unable to help themselves, no matter how many survival kits a Hercules aircraft might drop. The last-minute arrival of the helicopter made all the difference. And yet, even in this moment of triumph, the Labradors showed their advanced age. One of them was grounded for several days in St-Pierre because of mechanical problems. A replacement engine was flown in, repairs were made, and the chopper returned to its base at Gander. Another Labrador had problems with a leaky fuel-control unit and got stuck in Sydney as it awaited another engine. Eventually repaired, it returned to its Greenwood base. And so two of the four Labradors involved in the *Flare* rescue became unserviceable and unable to return to their bases without engine changes. Even with some of the best maintenance crews in the Canadian military, Labradors are extremely difficult to keep in the air. They have an availability rate of just sixty per cent, for example, and getting parts is a nightmare. The grounding in Sydney was especially ominous. That very aircraft, with tail number 305, was a ticking time bomb.

Recovery of debris from the crash of Swissair Flight 111. DND

Aircraft Down

Swissair Flight 111 and the Lessons of Chaos

Canada's East Coast lies beneath a busy air corridor for passenger jets flying between North America and Europe, their nine-kilometre-high contrails frequently visible on a clear afternoon. Every day, up to 650 flights cross through the region, most of them heading to or from the major population centres of the American seaboard. A much smaller number of short-haul domestic carriers adds to the aerial traffic flow. Only a tiny fraction of these flights gets into trouble while in transit anywhere along the corridor, compared with a much larger number of commercial ships in need of rescue or assistance, which speaks volumes about the high standards for aviation safety. And yet, every few years, one of these sleek engineering marvels comes crashing spectacularly to earth. And whereas the potential number of victims in ground search-and-rescue operations or bulk-carrier disasters typically ranges between one and perhaps three dozen, the stakes are usually much higher in emergencies involving passenger aircraft, which often carry a hundred or more people.

Just after noon on September 11, 1990, a Boeing 727 operated by the Peruvian carrier Faucett Airlines took off from Iceland's Keflavik airport after a refuelling stop. The forty-seven-metre three-engine aircraft had recently been leased to Air Malta and still carried the red-and-white markings of the Mediterranean airline. On board were eighteen people, three of them crew members and the rest Faucett employees and their family members, including children. Among those on board was the airline's chief pilot. Flight PLI976 was a so-called ferry flight, undertaken after the expiry of the Air Malta lease simply to return the aircraft to its base in Miami along a route unfamiliar to the pilots. The Boeing 727 was not originally designed for transatlantic flights but was intended rather for short-haul domestic work. This particular aircraft, built in 1969, lacked automatic navigation systems and a long-range radio, and its relatively small fuel tanks required the pilots to refuel after short hops. The next refuelling stop was scheduled for the international airport at Gander, Newfoundland, about three hours and twenty minutes flying time and more than 2,600 kilometres away. At Keflavik, the crew filed a standard "zero-wind" flight plan, prepared by a European aviation service company, that required the pilots themselves to factor in the winds along the corridor based on the latest weather reports.

Despite its origins in 1960s-era technology, the Boeing 727 is a tremendously successful and long-lived design. More than 1,800 came into service before the production lines were shut down in 1984. The aircraft is particularly suited to shorter runways, and it was attractive to the Peruvian airline because it could operate at high-altitude airports — it was the first jet ever to operate from Cuzco, Peru, for example, high in the Andes at 3,200 metres above sea level. The aircraft's sales record was surpassed only by that of its successor, the Boeing 737, and about 1,500 remained in service all over the world in 2001. Although Faucett's PLI976 was making an unusual transatlantic hop, the aircraft was capable, reliable and proven, and many smaller, more fragile aircraft are successfully ferried across the Atlantic each year.

The weather was good as the pilots left the bleak landscape of south-western Iceland for an easy cruise to Newfoundland. Forecasters estimated northerly winds along the flight path at about eighty-three kilometres an hour, a report the crew received while in flight. The steady pressure on the fuselage would nudge the aircraft south of its intended route unless the pilots compensated. The trip apparently proceeded uneventfully until the pilots reached the point where they should have been within radio range of Gander, in central Newfoundland, and should have seen some landmarks or lights. But air-traffic control could not be raised, and below them was the steel-grey North Atlantic, not the rugged, forested landscape of Newfoundland. Something was terribly wrong. Over the next hour, the crew contacted nine aircraft flying in the area to alert them to the problem and to ask that messages be relayed to Gander. The Boeing 727, the fuel tanks in its wings nearly drained, was clearly far south of its intended target. Pilots in the surrounding aircraft thought the crew was inexplicably slow to react to their navigation problem. The crew of a nearby American Airlines flight kept urging the Peruvian pilot to turn north. "It was getting him to turn around that was the problem," a Canadian investigator said later. "This American Airlines pilot had to tell him three times that he should probably turn around." The aircraft finally did turn north, but far too late. The pilot soon sent a distress call saying he was losing altitude with the low-fuel warning light on and was preparing to ditch in the ocean. Ten minutes later, the jet disappeared somewhere about 380 kilometres southeast of Cape Race, Newfoundland.

Over the next eight days, Canadian rescue aircraft and ships criss-crossed thousands of square kilometres of ocean with no clear idea of just where the jet had hit the icy water. There were dozens of debris sightings and reports of oil slicks during 610 hours of searching, but nothing could be definitively linked to the shattered Boeing 727. What appeared from the air to be a life raft turned out to be an old tarpaulin when a search ship finally recovered it. The captain of a Russian fishing trawler came across a cigar-shaped object twenty to twenty-five metres long and a metre high, painted in the colours of Air Malta. Though it was never recovered, rescue officials said they believed this was indeed a large piece of the broken fuselage. No bodies were ever found, and the aircraft's black boxes lay too far down in an uncertain stretch of ocean ever to be located, much less retrieved. There would be no easy answers to why a routine flight had gone so wrong.

Canadian investigators piecing together the weather data, flight plan and known radio communications later surmised that the pilots had failed to account for that steady north wind when determining their final flight path. A simple course calculation would have put the aircraft on the money in Gander. "He was off track right from the start," Major Bill MacDonald of the Halifax rescue co-ordination centre said later. "He wasn't a very smart pilot by the look of things. Nobody would expect a 727 to get lost. But he got lost."

Seven years later, another pilot error put many more lives in jeopardy. On December 16, 1997, a Canadair CL-65 regional jet approached the Fredericton airport shortly before midnight in a thick fog. As the twin-engine Air Canada aircraft was about eleven metres over the runway, it drifted too far to the left for a safe touchdown. The captain ordered the co-pilot to abort the landing and pull up to circle around for another attempt. The idling engines could not get back to full power quickly enough, and the aircraft began to stall as it veered to the right. Seconds later, the right wing-tip hit the runway, the nose-wheel broke off, and the now fully powered engines dragged the jet across a field and up a hill into a stand of trees, where it slammed into a fir tree. The trunk sliced into the fuselage, pinning some passengers in their seats and cutting all electrical power. Nine of the forty-two people aboard suffered serious injuries, and many of them had to pried or cut loose from the tangled fuselage over a period of more than two hours. One man later had a crushed leg amputated. Miraculously, no one died.

Canada's search-and-rescue system is responsible for covering large swaths of the country, with the notable exception of airports, which have their own emergency response personnel on hand and can also draw on municipal services, from hospitals to firefighters. But the ground crews at Fredericton airport were initially uncertain that there was in fact an emergency involving Air Canada Flight 646 because of the fog and the lack of any radio communications. Due to an archaic Transport Canada regulation, the aircraft was not required to carry a rescue beacon, designed to emit an emergency signal in a crash to alert rescue workers and to help them home in on the site. The airport's emergency crews could do little more than drive back and forth on the runway for almost fifteen minutes, looking for a possible crash. Some thought that the aircraft might simply be airborne, setting up for another attempt at a landing. At about two minutes past midnight, their worst fears were confirmed when a passenger walked out of the fog and into the dim headlights of an airport truck. "Somebody walking here on the runway," the maintenance foreman radioed to the airport tower. "He thinks everybody's okay." Rescue vehicles could not reach the wreck for another thirty-six minutes as they struggled through the deep snow.

Seventeen months after the Flight 646 disaster, the Transportation Safety Board of Canada delivered a final investigation report. It criticized Canadian regulations that allow landings to be made in such low-visibility conditions; airports in the United States, on the other hand, would have required the pilots to make another pass or head for a safer airport. The report also criticized the captain for permitting his relatively inexperienced and inadequately trained co-pilot to handle a landing in such difficult circumstances. Investigators found no mechanical problems or design flaws in the twenty-seven-metre jet.

A crash barely five months after the Air Canada disaster involved another instance of human error in the cockpit. A single-engine Pilatus PC-12 on a trip from St. John's to Goose Bay in Labrador got into trouble shortly after takeoff on May 18, 1998. Ten people were aboard the small, propeller-driven aircraft, including eight passengers. An oil-pressure light on the cockpit panel alerted the pilot to potential trouble, but he dismissed the warning as a false alarm because there had been a glitch with the indicator on at least two previous flights. The problem worsened, so twenty-three minutes after takeoff the pilot turned back toward St. John's to have maintenance workers examine the aircraft. The engine soon developed vibrations that seemed to subside as the pilot reduced power, but the shaking returned, forcing him to shut down the engine. The Pilatus suddenly went silent and began a long, horrifying glide. "Okay, sir, we have a complete engine failure. Complete engine failure. Repeat, complete engine failure," the pilot tensely radioed the Gander air-traffic-control centre. "I need vectors for the nearest suitable airport."

St. John's was no longer suitable. The airport was too far away for the aircraft to glide safely there from its current altitude of 3,900 metres, so Gander recommended the little-used airstrip at Clarenville, which sometimes doubled as a drag-racing track. The pilot turned north again toward Clarenville and continued his long glide, the view through his windshield obscured by spattered oil on the outside and by condensation on the inside. "If we can't make the strip, we may have to be on the road there so, ah, ah, just advise the RCMP there, I guess," the pilot told Gander. Some of the passengers were audibly praying during twenty agonizing minutes of descent, while one man opened his attaché case and began writing a farewell note to his wife. The aircraft finally broke through the cloud cover at an altitude of about 150 metres, and, with no airstrip visible from his side window, the pilot made for a bog, hoping the soggy, treeless patch would leave the fuselage more or less intact. Out of radar and radio range, they were just three kilometres short of the Clarenville runway.

The aircraft clipped the tops of four small spruce trees before it crossed over the bog, its left wingtip gouging a seventeen-metre-long trench before snapping off. The belly of the fuselage, its landing gear retracted, smacked the ground, and the engine was knocked free as the aircraft spun round before coming to a stop. The nose was a tangle of hanging metallic shreds, the windshields were missing, and only the right wing remained attached. During the crash, one of the passenger seats tore free of the floor rail and sent businessman Lloyd Hillier, forty-five, bouncing about the cabin. "We saw the end of the world, we were dead," he said later. The pilot and a company observer in the cockpit were left pinned in their seats with serious leg damage. All but one of the passengers escaped with only minor injuries, largely because the dripping fuel did not ignite on impact. One of the passengers took charge of the scene, organizing the exit of people from the cabin and firing a rescue flare. He also built a fire for warmth, using pages from the aircraft's flight manual and some of the spilled aviation fuel. An RCMP-chartered helicopter was first on the scene, and the pilot and his companion were meticulously extracted from the blood-spattered cockpit and sent to hospital.

An investigation by the Transportation Safety Board of Canada was not able to determine why the engine lost oil pressure but said the pilot should have headed immediately for St. John's at the first sign of trouble. Officials calculated that he would have had just enough engine life to make a controlled landing at St. John's, but his delay left him unable to reach any airstrip. Like the Air Canada regional jet that crashed at Fredericton, the Swiss-made Pilatus PC-12 did not have a rescue beacon; the battery-driven device had been removed for maintenance just before the flight. Transport Canada regulations allowed such removals for up to ninety days, but after the Clarenville crash the period was reduced to thirty days. Investigators said the lack of an emergency beacon did not delay the rescue.

Three crashes, three cases of pilot error. In each instance, the death and injury toll was mercifully limited because the Boeing 727 was nearly empty and the two belly landings resulted in only non-critical injuries. Investigators later determined that two of these three crashes may have happened because the pilots failed to head immediately to the nearest airport at the very first sign of trouble. Quick and decisive pilot action can make all the difference between a close call and a tragedy. Dozens of close calls go unreported each year because of sharp thinking in the cockpit. But once in a great while, pilots fail to respond swiftly enough to a relatively small incident that quickly grows beyond their ability to act, and a close call is transformed into horrifying disaster.

■ ■ ■

On the night of September 2, 1998, Swissair Flight 111 plunged almost vertically into the choppy waters off St. Margaret's Bay, Nova Scotia, crumpling like an empty beer can driven into a concrete floor. The force of the impact shattered the 200-tonne MD-11 aircraft — and the fragile bodies of the 229 people on board — into hundreds of thousands of pieces. The dark bay soon filled up with fishing vessels, coast guard boats and navy ships looking in vain for survivors. Rescue helicopters and planes buzzed overhead, their spotlights barely piercing the drizzle thrown up by Hurricane Danielle, passing 300 kilometres to the south. Although the so-called search-and-rescue mission would stretch to thirty-six hours, it was clear to everyone on the water after about four hours that there would be no rescues. All around, even in the black of night, was indisputable evidence of a catastrophic impact. Fishermen, hoping to haul chilled, frightened survivors from the water, unexpectedly encountered a grisly array of floating body parts. Men reaching over the gunwales to pull aboard an unidentified bobbing object would suddenly jump back, recognizing a human torso. Everywhere the stench of tonnes of jet fuel induced mild nausea. Plumes of debris rising from the ocean floor brought to the surface a horrifying pastiche of carry-on baggage, money, passports, postcards, stuffed toys and other detritus from violently shattered lives.

The story of Swissair Flight 111 has been told and retold many times, in newspapers, in books, on television, on film, in magazines and on Web sites. The official investigation into the cause of the disaster took years, though few doubted from the very first days that the aircraft experienced a wiring-related fire and massive electrical failure. The pilots were left powerless to control the aircraft after they turned away from Halifax International Airport to dump fuel, in retrospect an ill-advised course of action. The human tragedy for the families moved inexorably from shock to mourning and from grief to angry recriminations. The families' stories have been told, and will be told again and again as multimillion-dollar lawsuits consume the courts for years. Many of the stories of the rescue workers have also been told, from the first minutes after the crash to the months and years of nightmarish memories that must somehow be extinguished or tamed.

But one story behind the tragedy has yet to emerge, partly because it has been so well hidden behind a veil of bureaucratic discretion. Swissair Flight 111 had a profound and continuing impact on Canada's emergency organizations. Despite the near-heroic efforts of hundreds of men and women who made great sacrifices to help with the Swissair search and recovery, many of the organizations they worked for were ill-prepared for the magnitude of the disaster. Canada's emergency infrastructure was simply not ready for Swissair Flight 111, and only good luck camouflaged the often tattered response. "That is how it began, the first couple of hours. It was total chaos," Major Michel Brisebois, head of the Halifax rescue co-ordination centre, said at an internal briefing nine days after the crash. "Controlled, but chaos nonetheless."

The Swissair tragedy has triggered a minor aviation revolution. Pilots today react much more quickly and decisively to potential fires. Electrical systems and wiring have become subject to closer scrutiny and more frequent inspections. The effect can only make passenger jets, already a very safe mode of transport, even safer, offering some solace to grieving families. But Swissair also sparked a more hidden revolution. The crash was a wake-up call to officials in many of Canada's emergency response organizations, who found that their planning, their resources and their training were not up to the challenge. Fortunately, the disaster happened in a relatively accessible area, a three-hour steam from the navy's main East Coast base at Halifax and from the coast guard's main Maritime operating headquarters at Dartmouth. There were two military airports within thirty minutes flying time. Area telephone systems could readily absorb the sudden demand for communications, with the province's telephone company and maintenance staff headquartered in nearby Halifax. The area's cellular services were also quickly expanded for the crisis. Emergency medical services, put on high alert for the first few hours, were not tested at all, since there were no survivors to triage and transport to Halifax hospitals. Had the aircraft developed its electrical problems a mere thirty minutes later on its flight toward Geneva, and had it managed a more controlled descent into the North Atlantic, the system would have been instantly overwhelmed. Search and rescue after a mid-ocean aviation disaster with survivors would have been all but impossible to prosecute without loss of life.

Emergency officials knew almost from the start that they had gotten off lightly with the Swissair crash. And as the recovery operation dragged on for weeks, they began to throw out their decades-long assumptions about how to respond to major disasters. There has since been a fundamental rethinking about such basics as communications, command and control, training and equipment inventories. Indeed, the Swissair experience presented a rare window of opportunity as governments — suddenly alerted to the potential threat to their public image — became much more willing to open purse-strings and reverse a decade or more of funding cuts to emergency organizations. The great leaps forward in search and rescue have always come in the aftermath of great disasters, and the crash of Flight 111 turned out to be one of the most potent agents for change. The revolution is still underway.

■ ■ ■

Steve MacLeod was working alone the night of September 2, 1998, as the desk editor at the Halifax bureau of the Canadian Press. Things were extremely quiet in the small first-floor office on Brunswick Street. There was a light flow of news stories arriving electronically from a few of Atlantic Canada's fifteen daily newspapers, some of which MacLeod rewrote and sent out on the wire. There had also been Wednesday lottery numbers from Moncton, New Brunswick, that arrived by fax and had to be typed into the system for relay. At about ten p.m., MacLeod called a taxi to pick up the provincial edition of the Halifax *Chronicle-Herald*, the first copies of which had just rolled off the presses a few blocks away on Argyle Street. When the paper arrived, MacLeod scanned the contents to make sure CP Halifax had not missed any important stories. He was just preparing to send a routine note to the wire about its front-page stories and photographs when the telephone rang. On the other end of the line was a "tipster," one of dozens of people in the region who regularly watch for and tip the CP newsroom to breaking stories. Some of these people are government employees who do it as a public service to disseminate information quickly and widely. Others are members of the public who enjoy playing a small role at the periphery of the news industry.

The caller tonight had been listening to an electronic scanner that monitors police and emergency radio frequencies. "I'm probably going to ruin your night, but there appears to be a Swiss plane that just crashed," said the familiar voice on the other end of the line. "There's a fleet of ambulances heading out to Blandford." He said the scanner, which had been relatively quiet all night, had suddenly erupted with a flurry of urgent calls for emergency help in the St. Margaret's Bay area. There were hurried, vague references to a downed airliner. In the rarefied world of scanner-monitoring, a hobby akin to metal-detecting, this was the mother lode, the biggest find in more than a decade. MacLeod thanked the caller and immediately began dialling out to gather details on what sounded like a huge story.

In the news business, information gleaned from scanners is generally suspect and often misleading. Cops and ambulance drivers themselves often have only a few fuzzy details when trouble breaks out. Scanner information always needs to be confirmed and fleshed out with telephone calls to official sources or witnesses. After numerous checks, MacLeod finally gathered enough information to put something on the wire. A story of this potential magnitude called for a CP Alert, a brief headline-like declaration of an urgent breaking story. And so at 11:09 p.m. Halifax time, less than forty minutes after the crash, the Alert hit the wires: "BLANDFORD, N.S. — A Swissair jetliner has reportedly crashed in Nova Scotia." It was quickly followed by a story with a few meagre details gleaned from residents in the Blandford area, about forty-five kilometres to the southwest.

A CP Alert, always accompanied by bells or electronic squeals, commands attention in a newsroom. The CP service goes out to more than 100 daily newspapers across Canada and to more than 500 broadcast outlets, including the CTV and Global television networks and all branches of the CBC. MacLeod's brief story, initially sent on a trunk line directly to Toronto, was electronically rerouted to all these Canadian destinations. And because it involved Swissair, an international air carrier, the Alert and subsequent story was also routed to CP's American news-service ally, the Associated Press, in their New York offices. In an instant, then, MacLeod's story flashed across editors' terminals throughout North America and beyond. And within minutes, reporters around the world began to call into the Halifax rescue co-ordination centre for more details.

The rescue centre had been working on the Swissair case from the moment the Moncton Area Control Centre, the Maritimes' air traffic control hub run by the private agency Nav Canada, had called at 10:18 p.m. The message was that an MD-11 passenger jet from New York had declared a "pan, pan, pan" four minutes earlier because of smoke in the cockpit. Air Force Captain Mike Atkins was in charge of rescue centre operations during the night shift, only his third time in that top position. He was told the aircraft was diverting to Halifax International Airport for an emergency landing, where any rescue mission would be out of Atkins's immediate jurisdiction and a problem for airport emergency personnel. But Moncton called back at 10:32 p.m. to say they had lost the MD-11 on radar a minute earlier and had no more radio contact with the pilots. They provided the rescue centre with a compass bearing showing the aircraft's direction at the time it was lost on radar, as well as its distance from the Halifax airport area. But they did not provide an estimate of the longitude and latitude of the MD-11's last known position, a detail essential to building a precise search plan. And the rescue centre for some reason did not request the position. Instead, staff attempted to calculate the position themselves, a procedure that took an hour and a half because of uncertainty about just how Nav Canada had measured the distance, whether from the Halifax airport itself, a nearby radar site or some other reference point. "That's what created a bit of confusion . . . I'm not exactly 100 per cent sure during that evening whether it was simply the [rescue centre] controller at the time was too nervous or too engrossed in what was going on," recalled Major Brisebois. "Murphy's Law was working, and it just didn't flow very quickly."

Still, the rescue co-ordination centre had at least a ballpark estimate of where the aircraft was lost, off the mouth of St. Margaret's Bay about forty to forty-five kilometres southwest of Halifax. Atkins's team could only assume that the aircraft was somewhere between there and the Halifax airport; twenty per cent of this track was over water. At the same time, numerous 911 calls from residents in the area reported a low-flying aircraft followed by a loud bang. One caller, who had been sailing in the bay that night, advised that he saw an aircraft pass overhead with its lights on and a comet-like tail behind it, followed by a sound like a sonic boom. Another witness said there was fire on the wings. Whatever the discrepancies, the repeated reports suggested some kind of terrible disaster.

Convinced there had been a major air accident, Atkins's team ordered three coast guard vessels, the ships *Earl Grey* and *Mary Hichens* and the small cutter *Sambro*, to steam for St. Margaret's Bay. They would check the over-ocean section of the MD-11's presumed route to the airport, watching for the flaming jet fuel that had marked the July 1996 crash of TWA Flight 800 off Long Island, New York, which killed all 230 aboard. Meanwhile, a standby search-and-rescue Labrador helicopter and Hercules transport were ordered into the air from Greenwood to search the overland portion of the purported flight path. Other coast guard and navy ships were also called into the search, as well as a passing cruise ship, the MV *Veendam*, which offered a well-equipped medical facility with a doctor and three nurses. Just as the rescue centre was trying to co-ordinate the aircraft and the small flotilla and quickly develop a search plan, MacLeod's CP Alert hit the wires — and the place was suddenly inundated with telephone calls from reporters around the world.

"The media onslaught during the initial stages of this operation seriously hampered RCC's [the rescue co-ordination centre's] ability to do its job," says a post-operation report on the case. "The massive influx of telephone calls from the public and from media outlets around the world quickly impacted the RCC's capacity to communicate. For extended periods of time, the phones would continually ring in the Controller's hand as he/she hung up from the previous caller." Brisebois, who arrived at the centre soon after crash, found that even his personal cellular phone kept trilling with media calls. "There were limited opportunities to communicate with search resources, little access to free telephone lines, yet no incoming call could be ignored, as it could provide significant information like the discovery of survivors," the report continued. "To complicate matters, some members of the press even became abusive in their search for information." For an hour at the outset of the crisis, the centre lost control of the telephone system, its vital link to the outside world.

"I have twenty lines, basically, telephone lines — they were all taken by media," Brisebois told an internal briefing later. "I could not make a phone call to task an airplane to do something, and that bogged us down by about an hour." The problem was that for years the centre had informally allowed the media to call operational staff directly to check on whether any search-and-rescue cases were underway. Naturally, reporters called these same numbers the night of the Swissair crash and shared the dial-up information with others. At the same time, the telephone numbers at the centre followed an arithmetic sequence; it was a simple matter to call the next higher number in the sequence if the first one was busy. "The ability of the media to use 'call again' features ensured that as soon as a phone was hung up it would ring immediately," says a coast guard assessment. "There were no 'unpublished, non-sequential' phone lines that the controller could use to call out or to ensure that the air tower or the Emergency Measures Organization could call in." The effect was to freeze operations at the most crucial point in the disaster.

The confusion and tension also led to operational slips. The normally meticulous logs at the centre were not kept properly during the first few hours, a potentially serious deficiency because clear, timely entries reduce duplication and repetition. "Log keeping during the first few hours was very shaky," Brisebois said later. "Controllers were overwhelmed by the magnitude and pace of events." In order to get a clear picture of their own actions in the first few hours, staff had to play back and transcribe tape recordings of their numerous telephone calls to reconstruct events for the preparation of a final report on the operation.

In the aftermath of the 1982 *Ocean Ranger* sinking, the centre had developed detailed written procedures to deal with disasters at sea, but they proved useless that chaotic night. "We do have a major disaster plan," recalled Paul Rudden, in charge of the coast guard operations. "I pulled it out that night . . . and realized it is of absolutely no use to me right now. This is going to have to be something that we develop on the fly — and that's how it developed." Or, as the centre's final Swissair report put it, the plan "is a good familiarization and training tool. During this incident, however, it proved to be of little value as no one had time to read it as events unfolded." Oddly enough, Halifax's marine disaster plan had been trotted out as recently as June in a tabletop exercise to test the centre's response to a fictitious ferry sinking in the Cabot Strait separating Cape Breton Island and Newfoundland. Yet less than three months later,

Major Michel Brisebois at the Halifax rescue's co-ordination centre. DND

everyone seemed in the dark about its contents. Ultimately, in the recovery phase of the Swissair operation, the coast guard activated its Oil Spill National Response Team plan — the only model that seemed to fit the effort to collect debris and human remains from St. Margaret's Bay. Indeed, there was much discussion about whether to deploy giant floating booms designed to contain oil spills, except now they would be used to corral floating aircraft pieces and body parts. The provincial Health Department was likewise caught without an effective plan for the disaster. "Basically we built it from scratch," said Dr. Jeff Scott, a provincial medical officer. "We didn't have a plan, and that is an issue."

Nova Scotia's Emergency Measures Organization, responsible for co-ordinating air and ground ambulances, hospitals and other key services, seemed the most ill-equipped for the tragedy. Located in a suite of offices near Pier 21, the main entry point to Canada for generations of immigrants, EMO Nova Scotia was the only such outfit in the country without a proper emergency operations centre. Even neighbouring Prince Edward Island had three such facilities to deal with civil disasters. The province's four previous sites — in Kentville, Truro, Sydney and

Debert — had all been shut down in the 1990s at the end of the Cold War and the onset of government austerity. The organization now had only a modest boardroom with a false ceiling through which a few extra telephone lines could be dropped. One EMO official described it as a room "in which sixteen people are competing for twelve chairs in a room built for ten." Early in the search for survivors, the rescue co-ordination centre asked that EMO Nova Scotia send a staff member over to act as a liaison. "It broke our hearts early in the morning of September 3rd to have to say to the Rescue Coordination Centre, no, we can't send anybody because we don't have anybody to send," a Nova Scotia official said later. The shoestring organization had only seven Halifax employees and three others in the rest of the province, all of whom would be needed to staff the makeshift twenty-four-hour emergency centre in shifts. All contact with the rescue co-ordination centre would have to be by telephone, though with lines jammed at the outset of the emergency the agency had had to rely on ham radio operators for initial communications.

One early call from EMO that did get through — and that proved to be an annoying distraction to the bustling rescue centre — was a request for advice on where to get equipment to set up a morgue, including refrigerated trucks and body bags. "This request was unexpected, as it was assumed EMO would take the lead on this front, and detracted from our attempts to maintain operational focus," says the centre's post-operation report. The call, it turned out, was from a federal official working with the provincial agency; he was supposed to have called the local army headquarters. "It was a snafu," acknowledged the EMO chief, Mike Lester. The rescue centre then began to receive calls from the Nova Scotia Ambulance Service asking for updates on the situation off St. Margaret's Bay because EMO Nova Scotia was providing them with very little information. The centre, already overburdened, was forced to establish a direct liaison with the ambulance service to make up for the problems at EMO. The rescue centre's final report on the Swissair operation chided the provincial agency for communications that were "less than optimal, as EMO personnel and facilities seemed to be taxed with the magnitude of the required response."

Problems arose in the airspace around the St. Margaret's Bay area as well, which was declared off-limits for civilian aircraft but became dangerously crowded with government and military planes scanning the waves for survivors and debris. A Hercules transport and a Labrador helicopter, both from 413 Squadron in Greenwood, were first on the scene. Sea King helicopters from the air force facility in Shearwater joined the search later, along with a BO 105 light-duty coast guard helicopter, an Environment Canada DC-3, army Griffon helicopters and a Convair 580. However, the air force neglected to implement a proper air-traffic-control plan for the so-called exclusion zone, and flying was conducted without appropriate safety measures. Twice, Sea Kings and Griffons almost collided. At one point, three different types of aircraft wound up searching in the same grid until two were ordered out of the area when officials twigged to the danger. Only after ten days of intense searching did the military finally impose a localized air-traffic control system.

Sea King helicopter crews, meanwhile, chafed under a top-heavy command structure that had them answering to five different bosses to get their sorties approved. "The structure became five levels of control for one helicopter!" Major S.J. Newton, commander of the Sea Kings operating off the deck of the frigate HMCS *Halifax*, complained in a post-operation analysis. The situation was "inexcusable," Newton railed. "If a flight was to proceed without a hitch, it became necessary to call every level of control and pass on the same message, as well as talk to the aircrew." When communications inevitably broke down, "aircraft arrived without proper safety gear, timings were confused, and search tasking became one of personal preference [instead of] orchestrated logic."

The Swissair search and recovery effort also exposed the inability of separate government departments to talk to each other in the field. Only the RCMP, with its encrypted Saber radio system, had the advanced communications technology for such a complex operation. But the Mounties had only a limited supply of these units to lend to other agencies, and batteries inevitably ran down. In the end, cellular telephones saved the day. "Without cellular phones no one would have been able to talk to each other," Major Newton observed. "No other means of communication existed that would allow all government agencies to communicate effectively." Or as Don Bower, an EMO radio expert, put it, "We saw a brand new evolutionary use of cellular communication in emergency response. I think one of the more dramatic things I saw was a senior military officer setting aside his military communications equipment and opting to use a cellphone because it 'worked.'" The Red Cross alone brought in 120 cellphones, and several thousand were operating in the area at the peak of activities.

The problem with cellphones, though, is that they're vulnerable to eavesdropping, especially those using older analog technology. "Wireless service analog is easily scanned, regularly scanned, and people get their kicks off of that the same as they do police band or anything else," Eric Malloy, an emergency operations expert with Nova Scotia's telephone company, said later. Even digital cellphones, though less prone to eavesdropping, can still be scanned by a determined snoop, Malloy added. "It was very apparent that all cellphone communications were being monitored," said Jack Gallagher, then in charge of the coast guard's rescue service for the Maritimes. "At times decisions had to be made to breach security protocols in order to keep operations moving." The military found later that many of their own people were ignorant of just what communications equipment they possessed, causing the sometimes unnecessary distribution of a limited supply of cellphones and other devices. "On numerous occasions, additional communications equipment . . . was issued to units that were already holding identical equipment," says one report. The Internet also made its debut during the Swissair crisis as an ad hoc solution to pervasive inter-agency communications problems, even though it too is not always secure. "During this multi-national, multi-departmental operation, which included several different communication capabilities, the Internet became a primary source," the military concluded. "The route survey staff . . . and many others depended on the Internet as their primary communication link." The irony, of course, is that cellphones are the descendants of military-developed walkie-talkies, and the Internet was born as a Cold War military project. Civilian society had taken these tools, refined them to their full potential, and then bailed out their military forebears.

■ ■ ■

In a covered taxiway at Canadian Forces Base Trenton, Ontario, sit four pallets of specialized survival equipment and two Argo all-terrain vehicles. The fully enclosed space, which connects No. 7 and No. 8 hangars, is next door to Trenton's transport and rescue squadron and immediately adjacent to the Canadian Parachute Centre. Inside the taxiway, each self-contained pallet includes about 4,800 kilograms of emergency equipment and survival supplies. The contents, carefully packed in unpainted plywood boxes that can be broken up for fuel, include lanterns, stoves, tents, sleeping bags, generators, heaters, rations, water, flashlights, body bags and parkas. The whole load is 3.6 metres long, 2.7 metres high and 2.4 metres wide, larger than a minivan. The strapped-down pallet is designed to be ejected out of the open back ramp of a Hercules aircraft and floated to the ground on three thirty-metre cargo parachutes. Each pallet's kit can keep eighty people alive for three full days in the harshest, coldest environment. The two Argos, also rigged to be landed by parachute, are intended to haul all that gear by trailer directly to an emergency site. The four pallets and the Argos, a 100-person medical kit in Trenton and another in Winnipeg, and supplementary equipment in Comox, British Columbia, and Greenwood, Nova Scotia, constitute what's known as Majaid, the Major Air Disaster kit, which altogether can sustain up to 360 survivors of an aviation disaster in Canada's far north. Twelve members of the Canadian Parachute Centre, known as the "housekeeping platoon" and trained to quickly erect everything in the field, would be dropped along with the equipment.

Majaid is designed primarily for the High Arctic crash of a large passenger-carrying aircraft. "Passengers from a commercial airliner are generally not dressed for the en route environmental conditions to be found for much of the year in Canada's north," a military planning document notes dryly. The genesis of Majaid stretches back to the period after the Second World War when large passenger planes began to follow routes partly over Canada's vast north as they transited between Europe, Asia and North America. These flight paths appear circuitous on flat maps but are actually shorter because they hug the curvature of the earth. And because of Canada's geographical position vis-à-vis the United States, much of the rescue preparedness for these international flights falls on Ottawa's shoulders. In 1947, the federal cabinet formally approved funds to provide a rescue service to aircraft in distress, in accordance with recently signed international accords. An interdepartmental committee later refined this commitment to ensure that rescue aircraft "will be so equipped that emergency aid in the form of flotation, medical, exposure and survival equipment and supplies may be dropped or otherwise delivered to distressed persons in the interim period between the first location by the aircraft and ultimate rescue." The first modest air-disaster kits were developed in individual rescue units, but in 1965 the air force produced the first national contingency plans for a major crash north of sixty degrees latitude. The Majaid concept has since evolved to the centralized pallet system, which is unique in the world. The elaborate survival kit is a tacit acknowledgment that Canada is simply too vast and military resources too thin to expect quick rescue in the event of a major aircraft disaster. "Everyone else plans to pick up survivors right away," says Lieutenant-Colonel Charlie Cue, a search-and-rescue expert for the military. "We're the only country that doesn't do that."

Dumping almost 5,000 kilograms from the back end of a Hercules aircraft in one fell swoop requires a tactical crew specially trained in heavy airdrops. The goal is to get the first pallet into the air within six hours of a Majaid declaration. The process begins inside the covered taxiway at Trenton with a giant crane that hoists one pallet onto a flatbed loading truck. The vehicle then hauls the load out to the open ramp of a Hercules, raising it to the proper height. The pallet, specially designed with rollers and rails that fit into recessed tracks in the cargo compartment, is then winched inside the belly of the Hercules and secured with blocks. In the air, the cockpit crew uses a procedure known as the Calculated Air Release Point, or CARP, that combines data about wind speed, altitude, air speed and other variables to determine the precise moment when the load should be ejected. In the back of the aircraft, two loadmasters lower the back ramp during flight and deploy a 6.6-metre extraction chute that is attached to the pallet and that is designed to drag the load out the rear. On a signal from the cockpit, the blocks are knocked free and the chute hauls on the pallet, which rolls along the track toward the rear. As the massive load exits the aircraft, the green mushroom-shaped cargo chutes are automatically deployed, allowing the equipment to float gently to earth. Meanwhile, the sudden shift and departure of this enormous weight causes the Hercules to lurch, tossing passengers in the cargo hold. "If you're standing up, you might be kissing the wall," says one veteran who has witnessed the procedure. On the ground, honeycombed cardboard at the base of the pallet absorbs much of the shock of landing, and bubble wrap and other packing material keeps even the glass globes of Coleman lanterns from shattering.

The Canadian Forces employ two full-time reservists, both retired Sartechs, along with a part-timer to maintain all this equipment. Each main pallet, numbered M1 to M4, is torn down once each year to make sure nothing has become damp or mouldy and that mice haven't moved in. Food and water are refreshed, batteries and fuel are replaced, and everything is checked to make sure it's in working order. The contents of the four pallets are estimated to be worth about $535,000, the two Argo ATVs with trailers another $33,000, while annual maintenance adds about $17,000 excluding the costs of warehousing. International rescue exercises involving Majaid are held periodically with the Americans and Russians, driving up the kits' costs since everything has to be repacked and rerigged. And yet, despite the continuing commitment of military resources, in over half a century Majaid has never been air-dropped for a genuine mission.

Military planners estimate that the likelihood of a serious air crash in Canada will double, to two disasters a year, by about the year 2010. Currently, there are about 2.2 million flights by large passenger-carrying aircraft in or through all Canadian airspace annually, of which about 1.3 million neither begin nor end in Canada but are overflights between other countries. Worldwide, there are about 1.5 major crashes — that is, accidents in which the aircraft hull is broken — for every million departures. Tougher standards for airlines based in North America, though, have kept the rate on this continent to about one-third of global levels, or about one crash for every two million departures. Defence Department consultant Keith Gathercole says that although North American safety standards have risen in recent decades, the level appears to have reached a plateau, which means the accident rate is likely to stay the same over the next decade or so. Therefore, with 2.2 million flights a year, Canada can expect a little over one large aircraft accident annually. Statistics show that most crashes — about 95.5 per cent — occur when an aircraft is arriving or departing an airport or is climbing or descending. Just 4.5 per cent of such disasters happen in the "cruise phase" of flight, when an aircraft has reached its cruising altitude and is neither climbing nor descending. Crashes at or near airports are almost always primarily the responsibility of civil authorities, though military search-and-rescue resources are always available if needed. Majaid territory, though, is that other 4.5 per cent of accidents, when an aircraft is cruising or lost between airports. And the math tells military planners that such a catastrophic event will happen only every twenty years.

At the same time, airline deregulation, advancing technology and reduced barriers to the flow of people and goods around the globe have caused a huge increase in air traffic. Some projections see a doubling of flights as early as 2010, hence the prediction of two accidents a year in Canada. At the same time, non-North American carriers with less stringent safety standards may make up a greater share of flights through Canadian airspace, helping to drive up the accident rate. And with the end of the Cold War, the next few years will inevitably see a rapid increase in the number of polar flights across Canada and through Russian airspace. This route, which brings aircraft near the North Pole, can cut five hours off a seventeen-hour New York-Hong Kong flight and save about $50,000 in fuel costs. Or a Vancouver-New Delhi flight would take just thirteen and a half hours rather than the current eighteen hours, and trim the fuel bill by about $40,000. The first such flight from Canada occurred on May 19, 2000, when Cathay Pacific flew one of its Airbuses with 143 people aboard from Toronto to Hong Kong virtually across the North Pole. Nav Canada, the country's private air-traffic controller, plans to invest $7 million to facilitate polar routes, while its Russian counterpart plans a $33-million investment. Such flights represent a new challenge because the routes are over vast, unpopulated tracts with fewer emergency landing options and a difficult environment for survival. All these factors combined suggest that Majaid may now be called upon to rescue the survivors of a air disaster every five to ten years.

The military has only once had to cope with a large crash in the Arctic — and it was one of their own aircraft. On October 30, 1991, a Hercules transport bringing fuel to the remote military listening post of Alert on Ellesmere Island crashed after the pilot made a fatal navigation error. Four died immediately in the crash, just nineteen kilometres from Alert, but fourteen others lay stranded in the wind-whipped wreckage. Horrendous conditions, including fierce winds, whiteouts, extreme cold and twenty-two-hour darkness, prevented rescuers from getting to the crash site for two days. Majaid never got a real-life workout because it was never invoked. Instead, military resources were flown in from several southern bases as a kind of makeshift Majaid, considering there were only eighteen people aboard the aircraft rather than the more than 100 envisioned by the disaster plan. One of the biggest problems was getting rescue helicopters to the scene. Without air-refuelable or long-range rescue choppers, the choices were either to load one inside a Hercules transport or laboriously fly a Labrador north with frequent stops for refuelling at remote Arctic outposts. Both options were tried, both proved time-consuming, and the latter choice turned out to be highly dangerous.

Other long-standing equipment problems became life-threatening during the Alert rescue operation. Old-style mushroom-shaped parachutes based on Second World War designs gave the Sartechs little control over their descents. Many landed hard, sustaining minor injuries. The batteries in their hand-held radios, overly susceptible to cold, died within an hour on the ground, cutting communications. Equipment dropped by parachute was lost because, once it was down, high winds dragged the parachutes off with the vital supplies still attached; the Sartechs had not been issued canopy-release mechanisms that would have automatically cut loose the chutes. Strobe lights attached to equipment drops died within minutes because of inadequate batteries. Flare guns seized up after three or four shots because of extreme temperatures.

U.S. rescue personnel brought in to help with the disaster, on the other hand, were loaded with the latest equipment and technology. "The disturbing thread that runs through [the rescue operation] is that basic equipment deficiencies that have been identified for years on previous post-mission reports have yet to be corrected," said one military post-mortem on the mission. "A perception exists that search-and-rescue units are the last in line to acquire new technology." The rescue was especially embarrassing because search-and-rescue training had always been oriented toward civilian disasters, yet here the military couldn't even mount a timely operation to save its own people. The experience also showed that despite the best tabletop-and-shirtsleeve planning, the Arctic remains a formidable foe for rescue workers, and equipment located at southern bases may not be delivered to survivors until too late. Indeed, one of the five people who died in the Alert crash — Captain John Couch, the pilot — succumbed to exposure before rescuers could reach him. Boxtop 22, as the mission was dubbed, suggested that the problems of rescuing just thirteen survivors in the High Arctic stretched Canadian military resources nearly to the breaking point.

Majaid crews were put on operational alert for the first time ever in 1979, to respond to the disappearance of a Korean Air Lines Boeing 707 over the polar regions. But the mission was called off before any equipment was actually deployed when it was learned the pilot had set the aircraft down on a lake in Norway. The only other Majaid-level emergency on record occurred during the Swissair Flight 111 disaster. Major Michel Brisebois advised the National Defence Operations Centre in Ottawa at 11:30 p.m. local time, or about an hour after the MD-11 disappeared from Moncton's radar screens, that the Majaid kit might be needed. About fifteen minutes later, he made a formal request to the air force's operations centre in Winnipeg for the equipment and personnel. At the time, the crash site had not yet been pinpointed. Consultations, approval and detailed orders took another hour, though the necessary crews were immediately placed on standby. The first load of Majaid-related help — sixteen Sartechs and survival gear for 100 casualties — left Trenton in the belly of a Hercules soon after and arrived at the Shearwater airport.

Brisebois's first concern was to have enough medics at the two casualty reception points at Peggy's Cove and at Shearwater because EMO Nova Scotia simply did not seem prepared. "I wasn't sure how EMO was responding to this and whether we were going to have enough trained personnel to receive those kind of casualties," he later recalled. One of the four Majaid pallets was loaded onto another Hercules, but as the nature of the disaster became clearer, Majaid was stood down at 3:18 a.m. and no equipment left Trenton. Brisebois knew when he ordered in the Majaid kit that it wasn't a perfect fit, even if there had been survivors. "I certainly didn't need the tents and northern supplies, but I did need the medical supplies," he said. "I was looking for the field hospital, medical supplies, medical personnel."

And so Majaid — designed for Arctic air disasters with numerous survivors — has only ever been deployed, if partially, for a southern crash in late summer with no survivors. The tent heaters and parkas would have been excess baggage during the Swissair operation. "The Op[eration] Order that we dusted off the shelf for this, the Majaid, told us how to recover a Hercules aircraft in the Arctic but didn't apply very well to what we had with Swissair," as Captain Roger Girouard, a military planner, put it afterwards. And although Majaid was not invoked during the Boxtop 22 mission near Alert, that operation demonstrated how extreme weather and long flying distances can delay delivery of survival equipment beyond human endurance time. A contingency plan dating back over half a century would need to be revamped.

■ ■ ■

The search-and-rescue Hercules on standby at Greenwood got into the air an hour after the initial call from the Halifax rescue co-ordination centre about a passenger jet lost on radar. The crew was on scene twenty minutes later, scanning the peninsula to the east of St. Margaret's Bay. Its companion Labrador helicopter beat it to the site by just ten minutes. With nothing spotted over land, both aircraft were diverted to a broad area near the mouth of St. Margaret's Bay, the estimated spot where radar contact had been lost. Skies were overcast, and intermittent showers reduced visibility. The air was mild, though, about nineteen degrees Celsius, and the water temperature relatively benign at sixteen degrees — typical for late summer off this coast. Waves swelled to between one and two metres in height. The rain, from the coattails of Hurricane Danielle to the southeast, severely reduced the effectiveness of night-vision goggles. Even so, the Hercules crew soon spotted a concentration of fuel on the water and filed a Notice of Crash Location, or NOCL, officially confirming for the first time that the MD-11 had indeed crashed. The report came at 12:40 local time, or about two hours and nine minutes after the image had faded from Moncton's radar screens.

The information finally provided a finite point around which the Halifax rescue centre could calculate a reasonable search area that accounted for wind and drift.

The ocean now began to fill up with rescue craft, including eleven coast guard boats and many private vessels that joined the search without getting authorization from Halifax. Two navy ships, HMCS *Ville de Québec* and the supply ship HMCS *Preserver* — both of which had been doing sea training about fifty-five kilometres to the southeast — were on the scene in slightly over an hour. *Preserver*, with its cold storage lockers normally reserved for food supplies, was appointed the "on scene commander," in charge of co-ordinating all of the search boats — a total of thirty-seven by the end of the first day. The supply ship carried life rafts for 160 people and had a helicopter landing pad and a well-equipped sick bay with medical specialists. All surface vessels were asked to bring any survivors to *Preserver*, where they would be treated and transferred by helicopter either to nearby Peggy's Cove or to Shearwater airport, the two casualty-reception points being staffed by the Sartechs from Greenwood and Trenton.

The nauseating smell of aviation fuel was overwhelming at the estimated crash site, about seventy kilometres southwest of Halifax or nine kilometres southwest of the postcard-perfect fishing village of Peggy's Cove. The coast guard, still in its oil-spill response mode, later that day issued a health-hazard advisory that warned "high vapour concentrations irritate mucous membranes and lungs [and] may cause dizziness and headaches and be anesthetic . . . small amounts drawn into lungs from swallowing or vomiting may cause severe health effects." The kerosene-like vapours burned the eyes and throats of the search crews, and the filmy fuel residue melted surgical rubber gloves from their hands. A proposal to drop flares to help light the search area was quickly scuttled for fear of igniting the ocean, although the air crews from Greenwood had been dropping parachute flares through the night without incident.

In the absence of wreckage, there was still no clear sense of the scale of the disaster. But shattered debris and personal effects soon began to appear, bobbing on the swell: shoes, clothing, wallets, empty lifejackets, oxygen masks, small chunks of insulation, an orange-and-grey twenty-man life raft ripped to shreds, carry-on luggage, a self-inflating slide used for evacuating aircraft passengers, the aircraft's identity papers. The largest piece found overnight was about five metres long, but most of the debris was in tiny pieces, and only Styrofoam-backed chunks of aircraft aluminum made it to the surface. Much of it lined up along the tide rips stretching across the mouth of the bay. The coast guard ship *Earl Grey*, meanwhile, brought aboard a metal buoy with its tower cage sheared off and its base scratched, the fresh metal exposed, apparently in a tremendous impact at Northeast Shoal. Investigators later discounted rumours that the MD-11 had chopped it in two, but there was never a clear explanation for the buoy's damage.

Then, shortly before one o'clock in the morning, came the first reports of mangled body parts. The captain of a small fishing vessel said he had found a badly damaged body in the vicinity of the broken buoy. "Fishing vessel was withdrawn from search due to assessed severe emotional trauma associated with state of remains he discovered," the commander of *Preserver* reported to the Halifax rescue centre. As more human remains were identified by other craft, searchers quickly began to lose heart. "The most tragic aspect was looking and seeing that the reason no people were being found was that the impact had not only catastrophically affected the aircraft but the passengers as well," says a military report. "The fishermen of the area continued to help, but the horror was starting to sink in, and their numbers fell slightly as people grew too burdened by what they witnessed and returned to shore." The ship's doctor aboard *Preserver* examined the first partial body received and quickly determined that death had come by "extreme deceleration." This was the point — about three hours after the search had begun — when it became clear to insiders that there would be no rescues. Although the official declaration was still about thirty-three hours away, the operation was quietly transformed into a search and recovery mission. The Majaid response, capable of handling 360 survivors, was cancelled — only a miracle would produce even one survivor. A fleet of thirty ambulances that had stood ready all night at three staging points was told at about 4:15 a.m. that they would not be needed.

Preserver eventually collected 1,350 kilograms of human remains that first day on scene, including the body of a small child, and the crew packed all of it in casualty bags. "Coroner has stated that 1 body bag of human remains is not equal to 1 body," an entry in the Halifax rescue centre logbook notes grimly. At one point, a bulging body bag being hoisted aboard *Preserver* from the *Sambro* broke open, spilling its horrifying contents back onto the deck of the cutter. The high-speed impact had "degloved" the bodies, in the clinical language of coroners, that is, flesh was torn cleanly away from bone. Sea King helicopters carried all the material in sixty body bags to "B" or Bravo hangar at

The debris recovered from the crash of Swissair Flight 111 was carefully sorted and identified. DND

Shearwater, which was being hastily converted to a mass morgue. "Due to extreme dismemberment total numbers of bodies recovered cannot be determined except by medical examiner," *Preserver* reported. Even so, enough personal effects were plucked from the sea that the remains of twelve people could be positively identified aboard the vessel. The ship also collected 1,274 pieces of debris during the official search-and-rescue phase of the Swissair operation, while another 352 items were catalogued by RCMP officers on shore. The search zone ranged across 290 square kilometres, an area roughly mapped around a nineteen-kilometre radius from the crash position.

The search zone was calculated in part by employing the same CANSARP computer program used in the *Flare* disaster. Different variables for tide action, winds, currents and waves were punched in — including data about the position of the one relatively intact body that was recovered. A Hercules aircraft from Greenwood also dropped a total of four self-locating datum marker buoys, three of them set to mimic the drift of a person in the water by using a kind of sea anchor that catches currents. The fourth was set to duplicate the drift of a life raft. These buoys regularly transmitted their changing positions via satellite to a receiving station at the coast guard college near Sydney, Nova Scotia, which then relayed the data back to the Halifax rescue co-ordination centre. The buoy information helped to correct the CANSARP predictions, though one of the buoys was mistakenly retrieved by someone assuming it was Swissair wreckage. The CANSARP results generally predicted a steadily expanding drift of debris toward the west, onto the pristine shores of the Aspotogan Peninsula and the 365 islands that dot the area. Ground search-and-rescue workers, including members of the Halifax group, would eventually visit each and every one of these islands, as well as comb the beaches around Bayswater and Blandford, in the search for debris and human remains.

Amid the dark and confusion of the overnight search for survivors came the faint signals of an emergency radio beacon. The high-pitched continuous whine of the aircraft's emergency locator transmitter, or ELT, could be heard by a military Aurora aircraft overhead, by the coast guard ship *Mary Hichens*, by *Preserver*'s own radio operators and by a satellite relay system. The sound was unmistakable but far too weak for searchers to be able to home in on its precise location. The battery-powered device is designed to survive a crash and automatically emit signals to help rescuers quickly find the wreckage. The Swissair model ELT, manufactured in France, was fitted to the tail section and apparently popped free as designed at the moment of impact. However, it was never intended to float or even transmit in water, and it proved useless to the Swissair search in the first few hours. (The device was heard intermittently through the next week until the battery finally gave out.) Compounding the problem was the accidental triggering of another rescue beacon on a pleasure craft near Digby, Nova Scotia, that was being put away for the winter. "It confused the whole issue because we thought that that signal was part of the [Swissair] ELT, and within a few hours we found out it was a different one," Major Brisebois recalled later.

Calling off any search for survivors is a highly emotional event for the families of victims because it brings with it a terrible finality and extinguishment of hope. Governments are also wary of potential public backlash, especially in light of the TWA Flight 800 crash in which federal officials were castigated for their insensitivity in ending the search too soon. Canadian rescue officials always tread carefully when seeking approval to scale down or terminate a search. The shocking state of the Swissair victims' remains, which became clear within hours of the crash, left little doubt among search-and-rescue veterans that it was not possible for any human body to survive such a high-speed impact. With the eyes of the world on them, however, rescue officials had to play out the search scenario lest there be accusations later about any half-hearted efforts. But how to set a time frame to end an expensive, resource-depleting and somewhat risky search-and-rescue operation? Fortunately, Peter Tikuisis's software program — used for similar purposes in the MV *Flare* case — provided a credible answer.

Tikuisis's program — officially known as the Cold Exposure Survival Model — could predict just how long it would take for a survivor to be overcome by hypothermia in various environments. Theoretically, the mouth of St. Margaret's Bay in late summer offered some hope for survival. Ships in the area reported sea temperatures between sixteen and seventeen degrees Celsius, the warmest these North Atlantic waters get after two months of hot sun. Planners in the Halifax rescue co-ordination centre at first used Tikuisis's model for a worst-case scenario: how long would a moderately tired sixty-year-old woman, of average weight and clothed in a long-sleeved shirt, survive immersed in choppy waters at 16.4 degrees? The model predicted she would lose the ability to swim or tread water in one hour, the point at which her body temperature would sink to thirty-four degrees; in two hours and thirty-six minutes she would become unconscious from severe hypothermia, her internal organs having cooled to twenty-eight degrees. Death would soon follow. Such a survivor would have been dead before the first human remains were even reported by search vessels and aircraft.

The rescue centre also calculated the best-case scenario, a twenty-five-year-old well-rested male weighing about 210 pounds. Increase the water temperature to 16.9 degrees, and put a heavy sweater on him in addition to a long-sleeved shirt: the model predicted he would become incapacitated in twenty-nine and a half hours, unable to help himself, and unconscious in thirty-six hours from severe hypothermia. Here, then, was the absolute upper limit of the search, since no one could reasonably be expected to survive beyond this time. And so at 7:13 p.m. — about twenty-one hours after the search was launched — Brisebois sent a message to Defence Department headquarters asking that the operation be wound down the following morning at the thirty-six-hour mark. "After an extensive and thorough search of the area, and in light of the indications of a violent impact, it has been decided there is no possibility of finding survivors," said the request. The results of the cold-exposure prediction were also relayed to headquarters, which some hours later gave permission for a "search reduction" — a euphemism for an end to any active searching. After a day and a half, the Swissair operation would officially be a massive recovery effort, co-ordinated by the RCMP, which had launched a criminal investigation, and by the Transportation Safety Board of Canada, which had just been handed the most expensive and complex crash investigation in Canadian history. Search-and-rescue personnel would return to their normal watches while the army and ground search groups carried on with the unglamorous cleanup work.

The scale of the search for survivors dwarfed anything that has preceded it in Canada. The Halifax rescue co-ordination centre estimated that at least forty-seven boats took part, with an unknown number of fishing vessels and pleasure craft having joined without officially checking in. All told, the marine search consumed 622 hours of cruising throughout the search area by the known search vessels in the first thirty-six-hour period. Eleven military choppers skimmed the waters, along with six aircraft from Greenwood, for another ninety-nine hours of aerial searching. Use of the military aircraft and ships alone cost taxpayers about $1.7 million for the short search phase. The coast guard ships, RCMP officers and equipment, army staff and resources and other related costs would easily drive the final tally to $3 million or more for a day and a half of intense, exhausting and ultimately fruitless work. Volunteers, of course, worked for free, but their labour has been estimated as worth well in excess of a million dollars. Search and recovery efforts, which carried on for many weeks afterward, as well as investigation costs, would consume at least $60 million more.

■ ■ ■

One of the great strengths of the Canadian search-and-rescue system is the culture of the post-mortem. Every operation, no matter how minor or successful, whether military or civil, is reviewed with fresh eyes in the clear light of morning. Labrador and Hercules crews will spend an hour, if necessary, rehashing a mere training exercise to assess honestly where things went wrong and how they might be fixed. Major operational gaffes can be embarrassing in the glare of television lights, but they can also be tremendous learning opportunities for rescue teams courageous enough to admit their mistakes. Canada's search-and-rescue experience of the last decades, in fact, has been a series of leaps forward triggered by post-mortem analyses of disasters: the 1982 *Ocean Ranger* sinking, the 1986 search for Andy Warburton, the 1991 Boxtop 22 crash and others. The 1998 Swissair tragedy has joined this grim pantheon, sparking a fundamental rethinking of disaster-planning that is still under way and will be felt for years. The process, which has escaped media attention as questions focus on the cause of the MD-11 electrical fire, is salvaging something of value from an event that left hundreds of families grief-stricken and angry.

The most brutal assessment of the Swissair response came from the Canadian Coast Guard's top rescue planner for the Maritimes. Jack Gallagher's "lessons learned" report was leaked to the media by the Union of Canadian Transport Employees two days before the first anniversary of the crash. Gallagher blamed years of cuts in staff and ships for having damaged the service's ability to respond to major disasters such as Swissair. "Coast Guard, departmental senior management and the Minister must recognize that Coast Guard is poised on the brink of an abyss," he warned. "The Minister should be persuaded to inform Cabinet as to the situation so that they can decide whether or not to rebuild the lost capacity prior to a major failure and embarrassment." The staffing situation was particularly perilous. The Halifax rescue co-ordination centre had only six marine controllers who could take charge of sea searches. These trained professionals were normally spread over a seven-day week in twelve-hour shifts with a lot of forced overtime. They and their families had come to live with the constant threat of pagers that could squeal anytime. The Swissair case required three marine controllers — half the complement — to work simultaneously on fifteen- to sixteen-hour shifts, a situation that could be sustained only for two days or so. Had the fire aboard the MD-11 occurred much farther out to sea, the rescue centre would simply have been unable to staff a search that might have taken a week or more.

Overall, the coast guard dedicated more than 500 staff members to the Swissair disaster, which effectively shut down normal day-to-day business for weeks. The Dartmouth offices of the environmental response section were left vacant for three months, for example, and the work of two key advisory councils was suspended for the rest of the year. Major projects were left in limbo, courses were cancelled, telephones went unanswered. Ships' crews — including that of the *Sambro*, the first coast guard vessel on the scene — were placed under incredible strains. "The [*Sambro*] crew quickly became stressed and fatigued, with no standard procedure to call in the other crew to provide relief," Gallagher wrote. "This put undue stress on the duty crew and affected their decision-making capability and health." Yet senior management seemed to be blind to the impact of workforce reductions. "As we do not make great noise, there is a dangerous illusion created in some areas, that there is no impact on regular operations. Nothing could be further from the truth." In the end, Gallagher concluded, "our response was more costly and slower due to our lack of personnel."

The specific problem of insufficient staff at the rescue centre had already been the subject of a union grievance before the Swissair crash. The disaster, however, seemed to galvanize management, and in November a team was assembled to find solutions. Their detailed report, delivered in less than a month, was a shocking indictment of what the team called a "just in time" approach to providing search and rescue. Workloads had become crushing, fatigue was chronic, and there were simply not enough trained staff to run the place. The problem stemmed in part from advances in search-and-rescue technology. As recently as the 1970s, marine searches were typically restricted to daylight hours and fair weather. Night-vision goggles, infrared tracking devices and powerful radar equipment, however, were now piercing the fog, the rain and the night, making the rescue centre a truly twenty-four-hour operation. Rapidly evolving communications technologies had also begun to flood marine controllers with calls from ships, where in decades past communications were sparse and intermittent. New satellite communications had also expanded the scope of the Halifax rescue centre's operations to include much of the Western Hemisphere and beyond, even though its actual rescue zone was much smaller. Search-and-rescue cases now needed an average of six hours to resolve, or about two hours more than in the early 1990s, partly because of more false alarms. Indeed, just ten days before the Swissair crash, the rescue centre had spent more than six hours chasing down what turned out to be a false satellite alarm from an oil rig in the South China Sea. The new technology, credited with saving more human lives, was nevertheless taking its toll on the coast guard's own personnel.

The fact-finding report, along with the impetus of the Swissair search, gave the rescue centre the extra coast guard staff it badly needed. In July 2000, the minister announced that the number of marine controllers would be doubled to twelve, even though the workload team had recommended hiring only four more. The new controllers were all in place as of January 2001. In addition, six temporary radio specialists, who were to stay only as long as it took to implement a new satellite-based rescue system — that is, until October 1998 — would now become permanent employees at the centre. The enlarged personnel pool would at least help take the coast guard beyond "just in time" search and rescue. More staff were also added to the rescue centres in St. John's, Quebec City and Trenton, and new lifeboat stations were promised for eight locations in Atlantic Canada and the Gulf of St. Lawrence, part of a three-year investment of $115.5 million. The move appeared to be driven as well by the report of the Transportation Safety Board of Canada into the January 1998 sinking of the *Flare*, in which the coast guard was found to be spread too thin along the south coast of Newfoundland. Another $54 million was promised in 2000 to help upgrade the tattered coast guard fleet over three years, partly to improve search-and-rescue capabilities. The coast guard review of the Swissair case had also exposed the ineffectualness of the rescue centre's marine disaster plan, which everyone ignored because it was complex and inappropriate, having been written before much information was available in the field of multi-jurisdictional response to disasters. The Halifax centre's plan and the plans of the other two rescue centres in Trenton and Victoria were rewritten and harmonized in 2000 using a $66,000 grant.

Nova Scotia's Emergency Measures Organization, which had been severely criticized by the military for its tepid, amateurish response to the Swissair crisis, went through its own period of soul-searching. Unlike the federal government, which was awash in surpluses, the agency had to operate under a heavily indebted provincial government that was still in a program-chopping mode. The tiny office's $600,000 annual budget was effectively frozen, forcing make-do solutions to the staffing dilemma. In the end, agency head Mike Lester enlisted the province's eight deputy fire marshals, training them in emergency response procedures, and another four volunteers recently retired from the emergency preparedness field. Given another Swissair-scale disaster, Lester could now draw on more than twenty people to answer the telephones and liaise with the rescue centre. Protracted negotiations also finally re-established a Nova Scotia emergency operations centre, a direct result of the Swissair experience. The facility, opened in early 2001, is located alongside a Halifax municipal emergency centre and Emergency Preparedness Canada offices on the refurbished second floor of the main police station in Dartmouth. The new space is eight times larger than the old EMO offices. Plans call for computerized workstations with disaster-planning software that is standardized across Canada. In addition, Lester says his staff now has better information about where stockpiles of disaster supplies, such as body bags, are held.

The military plugged the holes in its own disaster-response capabilities fairly quickly. The rescue centre ordered new unlisted telephone lines to enable officers to get calls out even if reporters jammed the incoming lines. Plans call for a Web site that will post search updates for wide dissemination, and they provide for faster arrival of public-affairs officers who can run interference. The post-mortem on the crash uncovered another unforeseen kink in communications. In past incidents, the rescue centre enjoyed unrestricted access to air-traffic control information because the system was run by Transport Canada, a fellow government department. However, on November 1, 1996, the federal government turned over operations to the non-profit and privately run Nav Canada, raising fresh questions about jurisdiction. Nav Canada has a clearly defined relationship with the Transportation Safety Board of Canada, which includes full access by the board to quarantined air-traffic recordings, radar data and other key crash information. The rescue centre, however, had no formal agreements — and there were communications problems during Swissair despite Nav Canada's apparent willingness to supply data. "In spite of this free flow of information . . . none of the questions RCC Halifax asked could provide a detailed description of the aircraft's flight path as it dumped fuel on approach to Halifax," says the centre's post-operation analysis. Indeed, the rescue centre was not able to listen to the contents of the MD-11's last communications with air-traffic control until mid-afternoon on September 3, five hours after getting joint approval from the safety board and Nav Canada and more than half a day after the crash itself. The Halifax rescue centre has since acquired copies of a radar-data mapmaking program, allowing it to plug in raw radar data from Nav Canada to make an independent determination about a flight path without risking another communications snafu with air-traffic control. The centre has also formalized its right to Nav Canada data in a written agreement.

The most fundamental change wrought by Swissair, though, is in the military's major air disaster planning. "Swissair was a wake-up call to non-search-and-rescue people, who think air disasters don't happen in Canada," says Lieutenant-Colonel Charlie Cue, a Defence Department rescue official. The Majaid kit is an all-or-nothing Arctic-oriented response that is now seen as inflexible and inappropriate for the range of disasters Canadian search-and-rescue teams may face. The department is re-examining Majaid to make it more applicable to the broad spectrum of emergencies that the military is today being asked to manage, from Red River floods to central Canadian ice storms. With modifications, the equipment could also be used for train wrecks, ferry sinkings or any other civil disaster that results in mass casualties and survivors who need medical treatment, protection from the elements and evacuation. Majaid planners want the Majaid kit to be more segmented, so that the response to a non-northern disaster can readily exclude the parkas and tent-heaters now imbedded in the giant Trenton pallets. The 100-person medical kits held in Trenton and Winnipeg might also be broken down into smaller parcels.

Swissair demonstrated that many disasters will necessarily require multi-agency responses, including other government departments as well as private companies, such as airlines. Yet, the greater the number of players, the greater the potential for unnecessary overlap and duplication. Officials are examining how to integrate emergency equipment that large airlines stockpile anyway, such as the clothing, stretchers and body bags that Canadian Airlines kept at Vancouver airport. Future Majaid training exercises, traditionally the sole domain of the military, will ideally be carried out not only with airlines but with representatives from Foreign Affairs and the Health Department. The result can be more effective operations, from passenger identification and next-of-kin notification to the more prosaic co-ordination of hotel bookings and car rentals. Air Canada passenger and cargo jets could even be used to deliver some part of the Majaid kit as well as doctors and nurses. "Greater provision for the inclusion of these agencies in both the planning and execution stages of Majaid is a necessity, if only to avoid the inevitable conflict of interest and acrimonious public debate that commonly results when agencies perceive themselves to be working at cross-purposes," says a Defence Department report.

The military, meanwhile, is questioning whether it can any longer meet its six-hour target for Majaid delivery. Hercules crews specially trained in heavy tactical airdrops are increasingly hard to find as the air force sees more of its cockpit expertise bleed away to the lucrative private sector. The search-and-rescue squadron crews are already trained in the parachute delivery of lighter equipment, using a simpler but more accurate method to calculate drop points. These specialists may have to be given training in heavy tactical airdrops to help close the gap. There is still no solution in sight to the problem of getting rescue choppers to a remote location quickly. A larger consideration is whether to integrate the military's air disaster plan with the coast guard's marine disaster plan into a single document with a common checklist, as recommended by the Halifax rescue centre.

The Majaid problem also comes down to dollars: who will pay for the training, warehousing and deployment if a larger, all-purpose disaster plan is born from the Swissair experience? More cost-sharing among different levels of government is one possible solution. But an answer being heard more often in the corridors of government is a so-called "search-and-rescue tax" applied to every airline, train, ferry or cruise ship ticket — perhaps twenty-five cents — as a form of user-pay. Canada's search-and-rescue community has long resisted user fees out of fear that a person needing rescue may hesitate to call, compounding the difficulties of any eventual rescue. Canada's international marine and air agreements also make it mandatory to provide search-and-rescue services to anyone who may require them. But the Swissair disaster, among the costliest search-and-rescue operations in Canadian history, may be the catalyst for introducing the user-pay principle.

The thousands of men and women who dedicated days and weeks of their lives to the chaotic aftermath of the Swissair crash made Canada's search-and-rescue system function, despite its shortcomings. Elaborate off-the-shelf plans proved useless, communications were shaky and archaic, trained personnel were in perilously short supply, and resources were difficult to muster. But the gaps in the system were closed, thanks to the hard work and ingenuity of the front-line workers. As these professionals well know, mistakes always provide valuable learning opportunities and often are a rare occasion to pry dollars out of reluctant governments. The Swissair accident investigation will eventually lead to better designed aircraft. It has already brought about safer flying protocols. Parallel to the inquiry into airline safety has been a quiet re-evaluation of search-and-rescue procedures, resources and staffing levels that has already improved Canada's preparedness for the next airline disaster. Swissair was a reminder that it is not a question of whether a passenger jet will crash somewhere in Canada, only a question of when.

That Others May Live

The Life of a Sartech

Tim Eagle's day begins, as it often does, with his four-year-old twins Andrew and Emily bouncing merrily on his head at five-thirty in the morning. No danger of sleeping through an alarm in this household, though his wife Alannah and older son Daniel, nine, do stay tucked under the covers a little while longer. The lights in the big suburban home in Kingston, Nova Scotia, flick on, and soon the kids are being fed and getting dressed, and plans are discussed for the twins' imminent birthday party. By seven a.m., Eagle is on the road this overcast wintry morning, headed for the back entrance of the Greenwood air force base halfway up Nova Scotia's Annapolis Valley. It's mostly a pleasant country drive along some secondary roads. The rear entrance to the base, on the other hand, is ugly, with rusting oil storage tanks ranged on the left alongside some half-finished roadwork that scars the landscape. The car soon slips past a radar installation, silent hangars and an ammunition dump.

Eagle wears an orange beret, and his orange jumpsuit hugs his 209-pound frame. At thirty-nine, he looks extremely fit and healthy, though he'd like to shed a few annoying pounds. Balding slightly, with short-cropped hair, Eagle is starting to get a little grey at the temples. But he exudes energy and youth, especially when he smiles, which is often. He has an open, straightforward personality and a deliberate way of expressing himself — a man of action rather than words, a man born for the outdoors. He is also someone who clearly loves what he does for a living. Sergeant Tim Eagle rescues people, and he works hard at being ready to do just that whenever the call may come. Even his name seems perfectly matched to his unusual career as a military search-and-rescue technician, or Sartech: Eagle-eye Tim. To his colleagues, though, he's still plain old Tim. No nicknames have stuck, although he has acquired the middle initial "O" despite having no middle name. A military clerk insisted on filling every blank in an enrolment form and put "Other" as his middle name. So officially, it's Tim Other Eagle.

Eagle graduated in 1985 from the Canadian military's intense search-and-rescue technician course, becoming Sartech No. 325, that is, the 325th graduate since formal training first began in 1945. Only twelve people are normally accepted into this elite school each year, and usually at least one of them doesn't make the grade. The group is taught parachuting, diving, mountain-climbing, advanced emergency medicine, navigating, map-reading and ground-search theory. At one point each recruit is left alone in the bush to survive for four days on a handful of jelly-candy rations, Labrador tea and whatever he or she can scrounge from the land, such as berries or a snared rabbit. The brutal course, eight months long in Eagle's day, though now extended to almost a year, is among the most comprehensive in the world. The closest equivalent is the eighteen-month course given the 350 pararescue jumpers or "PJs" of the U.S. Air National Guard. Unlike Canada's 130 or so Sartechs, though, PJs have a decidedly military role, with roots going back to the Vietnam War. They're trained to resist interrogation, to handle a range of weaponry, including grenade-launchers and M16s, and to extract downed pilots from behind enemy lines — all in addition to their domestic rescue roles. The Sartechs, on the other hand, are almost entirely oriented to civilian rescue. A few are stationed at Canada's CF-18 fighter jet bases in Cold Lake, Alberta, and Bagotville, Quebec, to handle military aircrew rescues, and at Goose Bay, Labrador, to assist downed NATO pilots in training. But most are spread among four main search-and-rescue squadrons whose prime role is decidedly non-military. The only weapons they train with are rifles to keep marauding grizzlies and polar bears away from crash sites. When Canada's CF-18 jet fighters are sent for duty overseas, NATO personnel rather than Sartechs handle military search-and-rescue duties; Sartechs remain at home. In the rarefied and demanding world of civilian search-and-rescue, they are second to none.

The path that brought Eagle into this elite corps began shortly before graduation from high school in Galt, Ontario. "My dad said to me, 'What do you want to do when you get out of school?' I wanted to do outside work, not a desk job, and I saw a rescue guy hanging from a Sea King helicopter in Britain on a news clip. And I knew I wanted to do that, to see that expression on the face of the rescued person.' Eagle assumed this was coast guard work, but after some difficulty he discovered with dismay that in Canada it was actually a military career. He had no interest whatsoever

Sartech Sergeant Tim Eagle standing in front of a Labrador helicopter at CFB Greenwood, Nova Scotia. DB

in joining the Canadian Forces but was so determined to have a rescue career that he enlisted in 1979 to begin the long process of qualifying for admission to the Sartech course. This was a giant leap of faith, as fewer than one in ten qualified applicants got into the course in those days. Twelve weeks of basic training at Nova Scotia's Cornwallis base, Eagle says, "was the most horrible thing I ever did" — but he managed to get through, becoming soldier No. C68-760-389. Next, he applied for and was accepted into a medical training course at Canadian Forces Base Borden, Ontario, a move calculated to increase his chances of being selected as a Sartech recruit. After an additional month interning at an Ottawa hospital, he was attached to the airborne regiment at Petawawa, Ontario. And while on dive training with the navy in the Halifax area in 1985, he got word that his application to become a Sartech had been accepted.

Canadian Forces members cannot even apply to become Sartechs until they have served at least four years in the military and have attained the rank of corporal or better. In years past, before the radical downsizing of the military in the 1990s, several hundred applications would flood in annually for the dozen training spots available. Even in these leaner times, about 100 members request a chance each year to be squeezed through the Sartech wringer. Until 2000, the selection board in Ottawa picked the lucky twelve based solely on paper records, without administering physical or other tests. The procedure, however, was flawed; one or two recruits often flunked out in the first two weeks because they were in such poor physical condition, despite what the paperwork at headquarters may have indicated. "A guy might have a million courses on paper," one recruit was quoted as saying in a military newsletter. "He might look like a genius and can tell you the square root of a pickle jar, but needs help in getting it open . . . you can't tell that just by looking at his application form." Today, the board picks a couple of dozen candidates who are then field-tested in the winter at the search-and-rescue facility in Jarvis Lake, Alberta, before the final twelve are accepted. "The files on these guys, when they're doing a selection for Sartechs, are incredible," says Colonel Randy Price, a Labrador pilot who spent most of his career in search and rescue. "You would not believe the quality of individual that they have to pick from to narrow it down to just those who are selected to go on the course. So we don't have a bunch of gung-ho Neanderthals throwing themselves out the door of an airplane. The quality of people we're getting in the Sartech trade is light years ahead of twenty years ago."

Applicants — all non-commissioned — come primarily from the army, largely because of the intense physical demands of the job. Even before arriving at the Comox, British Columbia, base where the school is located, a recruit has to have demonstrated the ability to swim 675 metres in twenty minutes or less. And within the first week on course, each student must complete a sweat-inducing gamut of exercises in seventeen minutes or less: a 2.4-kilometre run, thirty-one push-ups, thirty-three sit-ups, eight chin-ups, rope-climbing and more. This gruelling physical hurdle — only one second chance is allowed, and within forty-eight hours — usually sees one or two heartbroken soldiers pack their bags for home. On the other hand, most of the rest are so fit that they complete the tests in fourteen minutes or less. Indeed, so motivated are the recruits that they often arrive with some dive training or parachuting experience acquired through their own resources, though that often means they have to unlearn some bad habits and non-standard procedures.

"It's very hard on the body," says Sergeant John Oakes, chief instructor at Comox. "The rejects are usually weeded out right at the start." Eagle recalls being nervous on course because so much was at stake. "If you took an injury, you were out of there," he says of the school's zero-tolerance attitude. The candidates who do thrive on course tend to be avid outdoorsmen, hunters and hikers familiar with compasses, knots and bush survival techniques. Some are what Oakes calls "adrenalin junkies" who feed off the excitement and intense pressure of a crisis. In recent years, the average age of Sartech recruits has been rising, from the early twenties to the late twenties and early thirties, as Canadian Forces downsizing has left a smaller pool of applicants, many of whom must rotate more frequently through overseas postings. This later start for entrants to the trade presents a special problem for the military. Highly trained Sartechs generally remain in the rescue business for rest of their military careers, but as individuals reach their mid-forties, two decades of bone-jarring, muscle-tearing rescue work takes its toll. Injuries are not uncommon, and most members are out by age fifty or have taken the handful of desk jobs available in the trade. Sartechs who start at age thirty may realistically have only about fifteen years of productive service ahead of them. The longest-serving Sartech, Chief Warrant Officer Don Lane, worked an astonishing thirty-one years before retiring in May 2000 at the age of fifty-three.

The intense physical demands of the job help to explain why this is overwhelmingly a male trade within the military. Although women made modest inroads in the profession in the 1950s — the so-called Para Belles, nurses trained in parachute-jumping — they were never regarded as equivalents because they were not subject to the same physical requirements. The first woman in modern times to be accepted into the course flunked the physical fitness test in 1988, as did the sole female recruit accepted into the 1999-2000 course. Indeed, only one woman has ever made the cut, Master Corporal Tammy Negraeff, a 1997 graduate now stationed at Comox and widely respected among her colleagues. Few women even apply to become Sartechs, and no wonder: basics, such as private toilet facilities aboard Labradors and Hercules, are simply not available. Apart from Negraeff, there are no role models to inspire women to take up the challenge, and the bonding and competitiveness that the military encourages to create team spirit also nurtures a distinctly male culture that a woman could understandably find inhospitable, despite the best intentions of colleagues.

■ ■ ■

At seven-fifteen in the morning, Tim Eagle unlocks the entrance to the Sartech hall on the first floor of the cluster of offices and meeting rooms attached to the modern hangar of 413 Squadron at the Greenwood base. The squadron was founded in 1941 at Stranraer, Scotland, during the Second World War, but ten months later it moved to Ceylon to help fight the war against Japan. It was in Ceylon that the squadron adopted a bull elephant with tusks as its emblem. As a result, when its Labrador helicopters or Hercules transports are airborne for training, they're dubbed "Tusker 29" or "Tusker 51" to identify their squadron of origin. The unit is officially a Transport and Rescue squadron, but military transport is a decidedly secondary role. Since 1968, the unit has been assigned search-and-rescue duties covering all of Eastern Canada and much of the North Atlantic. Among their many tasks is to provide search-and-rescue standby for the first thirty seconds following the launch of the U.S. Space Shuttle when it is bound for the international space station. Members also carry out frequent medical evacuations from remote northern sites inaccessible to provincial health resources. All told, the squadron devotes ninety-nine per cent of its efforts to rescue work and is probably the busiest search-and-rescue unit within the Canadian military. Numerous high-profile rescues often thrust 413's members into the media limelight, prompting others stationed at Greenwood to dub them the "Hollywood Squadron." They've also been called "413 Transport, Rescue and Maternity Squadron" after twins were born to a woman on a medical evacuation flight.

The glory is fleeting, however, and day-to-day operations are often routine. Eagle is the first to arrive this morning in the off-white Sartech hall, which is dominated by a narrow fifteen-metre-long table used to fold and pack parachutes. He puts on some coffee in an adjacent meeting room decorated with group pictures of Sartechs, flips on the computer at his cluttered desk in another small office, and immediately begins inspecting his rescue equipment. "That's six volts," Eagle mumbles, as he tests a medical monitor with a small green screen. Radios are next, as he checks battery power. And so on, through diving gauges, pharmaceuticals and parachute packs. As of eight a.m., just twenty-five minutes away, the squadron switches from a two-hour standby status to a thirty-minute posture. That means rescue crews must be airborne within half an hour of being assigned a mission by the Halifax rescue co-ordination centre. All of Eagle's mountain of gear must be in perfect working order so as not to delay takeoff.

Sartechs are equipment pigs. On any given day, they could be diving thirty metres into a brackish lake, parachuting into a tangled deciduous forest, dropping from a Labrador to the deck of a cargo ship in tossing seas, or injecting morphine into a badly injured hunter. The range of gear required to handle that spectrum of crises is extraordinarily wide, and the stuff is generally bulky and heavy. Much of it is stored in giant blue equipment lockers in the hall outside the Sartech room. Eagle and his colleagues have to use big orange-red carts with bicycle wheels to haul it all to the waiting Labrador or Hercules on the tarmac outside. While the pilots swiftly settle into their cockpits carrying little more than a clipboard and an overnight bag, the Sartechs are in constant motion in the back of the aircraft, hauling and stowing gear, securing straps, adjusting clips, double-checking that everything is on board and in working order. Every pocket in their jumpsuits seems to contain some small survival device, from a wide-blade knife to a compass and flashlight. They pack videocams and still cameras to record accident sites. An official air force manual estimates two Sartechs are responsible for about 270 kilograms of basic gear brought aboard a Labrador helicopter — including body bags, radios, rope and binoculars — and specifically warns them not to stow any unnecessary material. Like all search-and-rescue aircrew, Sartechs also carry military-issue American Express cards for unexpected hotel bills, meals and transportation that may arise from rescue missions that take them thousands of kilometres from home and occasionally across international borders.

At seven thirty-five, Eagle rolls his groaning equipment cart out into the well-lit hangar of 413 Squadron, which at this moment houses a grey-coloured Hercules and two bright-yellow Labrador helicopters. The Labradors are painted yellow because their only role is rescue, and they need to be highly visible to victims. The Hercules, on the other hand, have a secondary role in military transport in the squadron and may have to deliver troops and equipment in hostile conditions. Their grey paint is designed to reduce their visibility and make them less of a potential target. Eagle pushes his cart to the lowered back ramp of the Labrador with tail number 310, where he meets his Sartech partner for the day, Master Corporal Darrell Cattell. Like most of the Sartechs, Cattell is exceptionally trim and fit. The daily physical activity of the job simply does not allow much fat to accumulate. Eagle and Cattell both begin to stow their gear in the already crowded helicopter cabin while a ground technician wordlessly pokes a flashlight into the engine compartments above. The back ramp is tested by raising and lowering it. Everything seems to be in working order. "Looks like we've got two aircraft," says Eagle, referring to the two Labradors. "Two serviceable aircraft." This is a minor maintenance coup, since Labradors are typically out of commission about forty per cent of the time because of frequent repairs and maintenance. Any helicopter is a challenge to keep flying because of its complicated engineering, but the outdated Labradors are among the most difficult of

Master Warrant Officer Jim McCluskey assembles equipment in the cargo hold of a Hercules aircraft at CFB Greenwood, Nova Scotia. DB

the breed. Manufactured in the 1960s, Canada's Labradors have not been upgraded as aggressively as the Labrador fleets of other countries. Parts are increasingly hard to find and expensive to acquire, and the maintenance crews at 413 squadron must sometimes devise their own spares. Sartechs often garner the glory for high-profile rescues in Canada, but among the unsung heroes are the technical wizards who keep the squadron's aviation dinosaurs safely in the air.

Shortly before eight, with everything in readiness, Cattell and Eagle briskly mount the stairs to the second-floor office where the Labrador flight crew is already discussing the day's plans. The team as a whole is given wide latitude — they might decide to train over Prince Edward Island in the morning, have lunch in Cape Breton and do more training in Newfoundland in the afternoon. There is a palpable exhilaration in having powerful aircraft and vast geography at one's disposal. The Labrador remains on standby for any rescue calls throughout the day, but until a mission is declared the crew must pack in all the flight training it can. The Sartechs alone have at least twenty-four separate skills that they must exercise on a monthly, quarterly or annual basis. "Where do you want to go?" says Captain Doug Parker, the aircraft commander, who looks a bit like actor Michael Douglas. "We could go to Saint John. You have something to pick up in Gagetown?" Last month, Parker had taken two Sartechs out for night-jump training at the giant army base in Gagetown, New Brunswick, and they had left behind two night lights, used to illuminate the ground target. Quickly a plan jells: a flight toward Saint John, New Brunswick, for some training in conjunction with a coast guard vessel already on patrol in the Bay of Fundy; then a flight to Gagetown to retrieve the lights; then home to Greenwood by four o'clock, when the thirty-minute standby posture ends and Eagle's official workday winds down again. A technical glitch, however, unravels the arrangements minutes after they are formulated: the maintenance crew says the Labrador needs a short flight check with a skeleton crew before it can do any training. The check was supposed to have been made earlier, but there is no written record, and safety demands that the procedure be demonstrated as having been completed. Everything is put on hold for the moment.

The team now heads for the small auditorium, where the squadron holds a daily briefing at eight a.m. for all its members. In ten minutes, everyone is updated on the weather forecast for flying in the Atlantic region, on the status of all squadron aircraft, on any search-and-rescue missions overnight, on the lunch menu for the standby crews, and today a reminder about a base curling tournament. The mood, as always, is upbeat and friendly. Everyone in 413 Squadron seems to like his job, his colleagues and his larger purpose. Eagle quickly marches back to his computer to run off some schedules for a staffing meeting set for eight-thirty. As the chief Sartech scheduler, it's Eagle's unenviable task to ensure there are at least two Sartechs ready or on call twenty-four hours a day for rescue missions throughout the year, including holidays. In addition, he must calculate the impact of annual leave, off-base training and development and personal requests from the twenty-one shop members, of which Eagle ranks fifth from the top. The half-hour meeting today with air-crew schedulers is intense, as they try to meld aircraft maintenance and operations schedules with the availability of crew members. "We're going to be low on Sartechs as of the first of May," Eagle warns everybody. "We have guys going on course." A harried aircraft scheduler gives a similar prediction: "We're in the red for Hercs then."

The very nature of search-and-rescue work severely limits the personal freedom of search-and-rescue technicians, even beyond the restrictions ordinarily imposed by a military career. All on-duty Sartechs, for example, are required to wear pagers that are triggered by the Operations Room chief at the squadron whenever required. Family life can be severely strained by these unexpected rescue calls, and wives with jobs outside the home are forever coping with sudden child-care crises. The community of Sartech families is necessarily close-knit, and mothers often bail one another out when Dad is called away. Sartechs frequently miss birthday parties, anniversary celebrations and other key family events, thanks to the electrifying squeal of a pager. Drinking alcohol is strictly forbidden while on standby and for twelve hours before any scheduled flying — from bottle to throttle, as they say in the military. (This twelve-hour restriction, which also applies to all military pilots, is tougher than the eight-hour abstinence imposed by Transport Canada for civilian pilots.) Sartechs and other squadron members may not live beyond a certain radius of the Greenwood base — about sixteen kilometres — to ensure they can get to work quickly in a crisis. The geographical confinement often means limited housing choices or a requirement to live in private married quarters on base, for which Sartechs are given precedence. (The Annapolis Valley, though, offers well priced and attractive housing for those lucky enough to be posted to 413.) Training regularly takes Sartechs far away from their families, whether to their annual back-country survival treks, to the U.S. Navy diving facility near Key West, Florida, or to occasional national competitions with other rescue workers.

National training sessions and competitive events not only develop technical expertise but also build a vital camaraderie and pride in military occupation No. 131, as the Sartech trade is officially designated. Indeed, during the eleven-month initial training course, recruits are taught the history of the trade through daily trivia questions. On graduation, newly minted Sartechs are presented with a sterling silver coin stamped with their unique graduation number. Trade tradition requires that the coin be carried at all times because another member can issue a challenge by presenting his own — coining, as it's called. If a Sartech can't reciprocate by showing his coin immediately, he's required to buy a drink for the successful challenger. These challenges can occur anywhere and anytime, whether in the showers after work or on a practice dive. The rules have gradually been relaxed to allow a Sartech to keep his coin within fifteen metres, and some Sartechs order spare coins, at $40 each, for a measure of safety. If a Sartech should ever lose his coin and news of the loss becomes generally known within a squadron, he can face a devastating series of bar bills. The volunteer Para Rescue Association of Canada, formed in 1989, and its lively newsletter keep past and present members of the occupation in touch. There's also a brisk trade in embossed ball caps, sweatshirts and T-shirts in the squadrons, all to keep alive a strong sense of tradition and community and to help compensate for the many personal sacrifices. As graduates, the Sartechs have all taken the same oath, binding them like a secret society: "Without regard for my personal comfort or self advancement, to the best of my ability and to the limits of my physical and psychological endurance, I solemnly pledge to make every effort to return to safety those victims of disaster entrusted to my care by the assignment of the mission to which I have consented. These things I shall do that others may live." These last four words have become a motto for the profession.

■ ■ ■

Back in the Sartech shop, someone has flipped on an AM radio, which pumps out some moderately loud rock 'n' roll. Two army specialists are busily untangling the cat's cradle of lines for a reserve parachute prior to packing it. (They'll get a case of beer if a Sartech ever has to use an emergency reserve chute — provided it deploys properly.) One wall is dominated by a mural painted by an artistically inclined Sartech showing a Labrador carrying out a rescue hoist. Mounted on the opposite wall is a souvenir: an orange lifejacket from the fishing vessel *Veryan*, which sank in 1999 off Gaspé, Quebec. Five of the six crew were rescued by the squadron Sartechs. Eagle now is back in his office, fine-tuning the staff schedule with a complex computer program — yet another skill he has mastered. He talks enthusiastically about a pet project to create a search-and-rescue calendar using photos of dramatic rescues, some of which he pulls from a drawer. The colour shots show two miserable-looking Sartechs, their faces grubby with black fuel oil from a sea rescue. Then it's out to the tarmac, where Eagle loads yet more equipment aboard Labrador 310 prior to its pending test flight. The clouds have scattered, leaving the sky crisply sunny and cool. The endless pre-flight checks and rechecks drag on through the morning until it becomes clear there won't be much flying time for training today. The Labrador finally gets airborne for a brief test at twelve-thirty and receives a green light for one afternoon training flight.

Meanwhile, the entire five-man crew, including Eagle and Cattell, pile into a squadron van for a bumpy ride across the base to the cafeteria for lunch. Although the three flight crew are all officers and the two Sartechs are enlisted men, there is little evidence of a caste system within the search-and-rescue world. The banter is non-stop, friendly and inclusive. The camaraderie is especially strong on cramped helicopters, where the flight crew typically drops off Sartechs on a mission and then picks them up again, thus working as a team together throughout an operation. Even on Hercules missions, in which the flight crew is physically separate in a remote cockpit and Sartechs cannot be retrieved once they parachute out, there is a powerful bond among all crew members, whether officers or enlisted men.

Partly the bond stems from the very nature of military search and rescue. Many members of the Canadian Forces share a latent hostility toward the rescue specialization, which is primarily focussed on saving civilian lives. The military's core mission is to be ready to fight a war, the traditionalists argue, and anything that does not directly support that goal is peripheral. For years, helicopter pilots who expressed a desire to do search and rescue were warned that it could be a career killer. And until the 1980s, air force promotions were unheard of within the specialization. The climate has changed somewhat in the last decade, but within the air force, no search-and-rescue specialist has yet attained a rank higher than lieutenant-colonel. Anyone wanting to move to colonel and beyond must get out of the business. Search and rescue thus remains a kind of ghetto within the military that — paradoxically — helps to build lasting friendships among those who have chosen to ignore career advice and adopt the specialization.

Another factor that binds these crews is the shared gratification from successful rescues. "It's like a video lottery terminal," says Colonel Randy Price, former commander of 413 Squadron. "You immediately get extreme and positive reward, and you get hooked on it. . . . In the military, you train all your life for something that you hope never happens. Nobody in the military wants to go to war, not at the war-fighting level. But in search and rescue, you train for something that you do every day. And when you go in and it works, and you save somebody's life, and you come out, you're on top of the world." At the same time, dangerous missions often mean putting one's life in another crew member's hands — a situation that discourages pettiness and backbiting. The banter among search-and-rescue crews, whether in the air, at lunch in the base cafeteria, in the Ops Room or at curling matches, is almost always friendly, light-hearted and mutually supportive. Lone wolves have no place in this culture and are quickly weeded out. Off-duty Sartechs often show up at the squadron to help their on-duty buddies. And despite the fact that search-and-rescue units are overwhelmingly based in English Canada, there is a striking mix of anglophone and francophone among the personnel, and language differences do not appear to create cultural rifts. Most Sartech shops have a strong complement of Quebec natives, and it is not unusual to hear a mix of French and English in the cockpits and passenger cabins of search-and-rescue aircraft. Typically, a rescue team can offer bilingual services to victims — adding to their uniqueness among the world's elite rescue personnel.

Eagle and the other men are back on the tarmac at one-thirty. The day's plan now has been scaled back significantly: the crew will rendezvous with the *Cumella*, the Canadian Coast Guard vessel already on patrol off Saint John. Once overhead, the Sartechs will carry out some hoist training to keep their skills sharp. Labrador 310 is now designated Tusker 51 as everyone climbs aboard for takeoff. The twin rotors begin to spin rapidly, creating some nasty vibrations inside the cabin that shake the loose grey padding of the ceiling and walls. The air outside is relatively calm. Strong winds can play havoc with floppy rotors that are revving up to speed, sometimes preventing takeoff altogether. Other times, the crew must temporarily move away from the section of the fuselage that would be sliced by an errant blade. Today, though, the rotors smoothly reach their pre-takeoff speeds. The sound is deafening, even inside snug-fitting air force helmets. Most of Eagle's and Cattell's equipment is strapped firmly to a stack of three metal stretchers fitted to the left side of the cramped interior. Eagle and Cattell are themselves belted into their chairs on each side of the cabin, about three metres back from the cockpit bulkhead. At 1:47 p.m., the aircraft taxis to its takeoff point about 100 metres away and is airborne ten minutes later, heading west. Skimming over the Bay of Fundy at about 230 metres, Tusker 51 tries to raise the CCGS *Cumella* on the FM marine radio with no success, though there's no problem reaching the coast guard station in Saint John. Eagle and Cattell, meanwhile, get dressed in body harnesses and haul equipment for the hoisting exercise to come.

At two-thirty, the pilots have finally contacted the ship, which is about five kilometres away. As the bobbing speck resolves itself into a red vessel, the Labrador's side hatch is opened, the hoist is readied, and a Stokes litter — a kind of stretcher — is placed by the door. The whole crew orally rehearses the emergency procedures should a mechanical failure dunk the helicopter into the sea. The *Cumella* is now rolling vigorously in heavy seas below as the Labrador hangs back and to the left, the pilot in the right-hand seat keeping the ship steadily in view. Eagle connects his body harness to the hoist cable and prepares to leave the rattling aircraft. The co-pilot calls out the changing altitude to the pilot as the flight engineer tells him the distance he should fly to position the Labrador over the *Cumella*. At the right moment, Eagle exits the aircraft with a slight jump, and the flight engineer carefully but quickly lowers him to the pitching deck, using a remote that controls the high-speed hoist bolted to the outside of the aircraft. Over the helicopter intercom system, the flight engineer calls out Eagle's declining altitude as he descends — fifty feet, twenty-five, fifteen, ten feet. The crew is tense in the rough conditions, and there is a sense of relief when Eagle's feet finally touch on deck and he steadies himself. The flight engineer plays out just enough cable so as not to accidentally jerk Eagle into the sea as the ship lurches, but not so much that it might wrap around his leg. No ordinary career this: printing out computer-generated work schedules in the morning, hanging off a thin cable over the snarling Atlantic in the afternoon.

■ ■ ■

The roots of the Sartech trade reach back to 1943, when two aircraft mechanics based in Edmonton were sent for parachute instruction to a U.S. Forest Service school in Missoula, Montana. Owen Hargreaves and Scotty Thompson, both civilians, were employed at a training facility run by Canadian Airways Training Ltd. to produce air observers and navigators for the war effort. Their boss was veteran bush pilot Wilfred

Sergeant Tim Eagle after being lowered to the deck of CCGS *Cumella*. DB

May, who had watched with dismay the growing number of air crashes occurring in the heavily travelled corridor from Edmonton to Alaska and beyond. Several thousand newly manufactured fighters and bombers were ferried along this route to join the war against Japan, many of them destined as aid-in-kind to the Soviet Union. The fresh-faced, inexperienced pilots assigned to deliver these aircraft frequently got into trouble and were forced to land in isolated, rugged terrain. Too often, May found, rescue flights arrived long after the young pilots had died of their injuries or succumbed to the elements. So he decided to organize his own team of rescue specialists who could quickly parachute to a crash site, administer rudimentary first aid, and perhaps save some young lives.

In 1939, the U.S. Forest Service had pioneered the use of Smoke Jumpers, parachute-trained firefighters who could quickly attack bush fires before they became a serious threat. The success of the service's school soon attracted the attention of the U.S. Army, the U.S. Coast Guard — and of May — who saw the potential for rescue work. May sent Hargreaves and Thompson down to Montana for the intense six-week course, where they eventually carried out about a dozen training jumps each. The pair were introduced to specialized equipment designed to safeguard a jumper coming down into trees: a high-collar jacket to protect the neck from branches (still in use today), a harness easily detached in case the jumper got hung up in high branches, and a leather football helmet with a wire-mesh face protector. On graduation, Hargreaves and Thompson brought four sets of the equipment back and themselves trained two more men at the Edmonton school to form the first "para rescue" team, the ancestors of every modern Sartech.

May managed to persuade the Royal Canadian Air Force of the necessity for specially trained air-rescue personnel. So in 1944, three members of that first team — the fourth had died in an unrelated accident — enlisted in the RCAF, and this is the date that Sartechs cite as the birth of their trade. The three were soon conscripted to help teach the first official military course, which consisted of nineteen weeks of parachute training, bush survival, first aid and mountain climbing. Some 20,000 men were said to have applied for the course, a list that was eventually whittled down to just twelve. The air force advertised for fit, "temperamentally stable" men between the ages of twenty-two and thirty, between 135 and 165 pounds, and with preference given to those with experience in the bush. (Age and weight restrictions have long since disappeared.) Course No. 1 began on February 12, 1945, in Edmonton and moved to a nearby wilderness area as well as to Jasper, in the Rocky

Mountains, where modern Sartech training is still carried out. All twelve graduated, and ten more enrolled in a second course held that same year. Course No. 5, offered in 1951, graduated four "nursing sisters," the first women to work in the profession. Three more were trained in the 1952 course before air force brass decided that staying current with parachute jumping interfered too much with the nurses' need for continuous medical training. The Para Belles faded into history.

Sartechs, previously restricted to jumping from fixed-wing aircraft, began to parachute from helicopters in the 1950s as rotor-wing technology advanced. The period from 1960 to 1975 is regarded as their most hectic: with a complement of fewer than fifty men to cover the whole country, the Sartechs were kept extremely busy as civil aviation and marine shipping expanded rapidly. Emergency locator transmitters were rare, and rescue crews had to visually scan large search areas for days at a time. In 1975, the military added dive training to the complement of skills Sartechs must master. Four years later, these "rescue specialists" — their official designation — became members of a formalized military trade with a new title: Search and Rescue Technician. All are now under the air force umbrella. In 1998, their medical training was significantly upgraded and includes time spent in hospital emergency wards helping to treat accident victims. Today, they operate under a standing order that allows them to administer morphine to a patient without a doctor's prior approval. Their standard medical kit includes an umbilical cord clamp and forceps for difficult childbirths. They carry a small portable pharmacy, including diazepam, nitroglycerin, codeine and tetracycline — a change from wartime days, when medical rescue kits included cigarettes. Thus over fifty years, while the rest of the military has witnessed increasing divisions of labour, the Sartechs have become jacks-of-all-trades, among the most broadly trained specialists in the rescue world.

Equipment upgrades arrived much more slowly over this same period, often leaving the Sartechs with outdated, inadequate gear that was frequently life-threatening. Penny-pinching military brass rarely acted on the men's own pleas for new kit but were goaded into spending only when a high-profile mission provoked the ire of the public and politicians. One such case was the search for Marten Hartwell, who crashed his Beech 18 aircraft on November 8, 1972, while carrying out a medical evacuation from the Arctic community of Cambridge Bay to Yellowknife. On board was a British nurse, a fourteen-year-old Inuit boy and a pregnant Inuit woman. The massive search effort stretched over four grim weeks with no sign of the aircraft or any survivors. A Hercules on a military supply flight

"Spotting" through the Plexiglas SAR door of a Hercules aircraft. DB

finally picked up an emergency radio beacon on December 7 and soon rescuers were scrambling over the snowy ground to the crash site. There they were welcomed by the only survivor, Hartwell himself, who told his startled visitors, "Welcome to the camp of the cannibal." Hartwell had managed to survive by eating most of the flesh of the nurse, who had died in the crash.

The crash site was well outside the search area defined by the military, and questions were soon raised about the effectiveness of the search-and-rescue operation — especially when it was learned that the Inuit boy had survived twenty-three days before succumbing. The result of the affair was to open the purse-strings in Ottawa for the improvement of search-and-rescue equipment. One of the first projects was to develop better conditions for "spotters" aboard the Hercules. Spotters, or observers, work in twenty- to thirty-minute shifts, watching the ground for tell-tale signs of a crash or marine disaster. At the time of the Hartwell crash, spotters simply lay strapped onto the floor of the open Hercules ramp, staring down as fierce Arctic winds drove temperatures inside the fuselage far below zero. Military designers experimented with a plywood and Plexiglas "spotter's box" that afforded some protection on the open ramp before they finally settled on a Plexiglas "SAR window" that can quickly be installed in the openings on both sides of the tail of the aircraft, openings normally used to deploy paratroops. Spotting can therefore be carried out with the ramp closed, and heat can be pumped into the cargo bay. SAR windows, still in use today, greatly improved the quality of life for the Sartechs who frequently act as spotters on long searches. The Hartwell crash also prompted development of a national search-and-rescue manual to standardize rescue procedures across the country; the installation of aircraft homing equipment, allowing crews to quickly pinpoint the source of emergency transmissions; modification of the military's Twin Otter aircraft to allow for unplanned parachute jumps; technical upgrades and new toilets for the Buffalo search aircraft; and the requirement, by 1974, for most civilian aircraft to carry the rescue beacons known as emergency locator transmitters or ELTs.

But the changes wrought by the Hartwell search were minor compared with the reforms following Boxtop 22, the 1991 crash of a military Hercules aircraft just outside the remote listening post of Alert, the most northerly community in the world. That difficult two-day rescue operation exposed a wide range of problems with the outdated kit the Sartechs had struggled with for years. Old-style parachutes, inadequate radio and strobe-light batteries, equipment parachutes without release mechanisms, flare guns that seized up — all these contrasted starkly with the leading-edge equipment used by U.S. rescue personnel on the same operation. In the aftermath of the mission, embarrassed Canadian Forces officials suddenly found money for a long list of requests that had been gathering dust at Defence Department headquarters since as far back as 1980. New gear came flooding into the search-and-rescue squadrons: night-vision goggles, Global Positioning System units for navigation, square CSAR4 parachutes, new signal-flare guns, better radios, and canopy-release mechanisms for air-dropped rescue supplies.

Today, Sartechs are relatively well equipped — apart from the geriatric Labradors they must work from — and their training is second to none in the world. Many do harbour grievances about their pay levels, however, especially since the massive cuts to defence spending in the 1990s. Most Sartechs are reluctant to speak publicly about their salary problems, but in 1998 they made a presentation to a parliamentary committee, complaining that despite their dangerous profession they don't even make as much as an auto worker. "For the sort of work that I do, I believe I should earn as much as a guy working on an assembly line making cars," Warrant Officer Jim McCluskey said following his appearance before the committee in Halifax. A newly graduated Sartech corporal, for example, makes about $47,000 a year, which includes a $4,500 rescue-specialist allowance that is not counted when determining a pension. The highest-ranking Sartech, with eighteen years or more in the profession, can make about $71,000. Most, however, make about $55,000 for doing one of the most health- and life-threatening jobs in the world. Sartechs, for example, must be vaccinated against hepatitis A and B viruses to protect against infection at emergency sites, and they never really know what germs they're exposed to during a rescue. Though their pay levels have improved slightly, consistent with general increases for all Canadian Forces members, the military nevertheless has turned aside their requests for a higher pay grade and for allowances that would take into consideration their dive training, aircrew status and parachute expertise.

■ ■ ■

Tim Eagle steadies himself on the deck of the heaving *Cumella* and deftly ties a rope to the end of the cable hoist. As the cable is winched back at ninety metres per minute into the hovering Labrador, the rope is hauled up as well, connecting aircraft and boat in this strange air-marine ballet. Eagle's partner, Darryl Cattell, is next to descend on the cable, but this time he is guided to the boat by the wind-whipped rope, which Eagle coils on the deck as the cable lengthens. The flight engineer again calls out the distances to the pilot as Cattell drops quickly through thirty metres or so before he is grabbed on deck by Eagle. Next, a Stokes litter is sent down to the *Cumella* twice to practise the extraction of an immobile victim, again guided by Eagle's rope. Then a rescue sling, known as a horse collar, is lowered, and both Sartechs come up together, one in the sling, to mimic the quick rescue of a survivor from a marine disaster, as in the *Flare* operation. With everyone back inside the Labrador, the side hatch is closed, and the pilots head back to Greenwood with a radioed thank you to the *Cumella*'s captain for the training help. A deck hand, who has been videotaping the whole exercise, waves goodbye from below. The training sequence has taken less than thirty minutes. Equipment is soon stowed, and the crew decides to do a debrief right away rather than wait until after landing.

Every training day at 413 Squadron is followed by a post-mortem among the aircrew, where all can talk frankly and openly about any problems or missteps and about ways to improve performance. These sessions, usually in the quiet of a meeting room, are often brutally honest though never vindictive. The *Cumella* exercise has been textbook, however, and there are no serious criticisms. "I think it went really smooth," says Eagle as he tucks himself into his seat. Half an hour later, the Labrador is on the ground outside 413 Squadron's hangar, and Eagle and Cattell are unloading their gear by the cartful. The thirty-minute standby status ends with a public address system announcement at four o'clock: "Standby crews are released, standby crews are released." Two fresh Sartechs attached to pagers are now on call for two-hour standby for the next sixteen hours. Eagle, though, still has an hour of gear-stowing and paperwork ahead of him. He finally pulls into his driveway in Kingston at five past five in plenty of time for horsing around with the kids and supper with all the family. If anyone at the dinner table asks what he did at work today, he'll probably skip over the meetings and scheduling routine and tell everyone all about dangling from a helicopter over the Atlantic Ocean.

Labrador helicopter No. 306 at CFB Greenwood, Nova Scotia. DB

CHAPTER FIVE

The Aging Rescue Choppers

The Final Flight of Labrador 305

The final few seconds of the last flight of Labrador helicopter 305 were so very complex, so riven by disaster, that to understand the terrifying sequence you have to expand time.

The passenger cabin of a Labrador is long and relatively narrow, with a loading ramp at the rear that is raised shut for flight. On the right-hand side of the interior is a bench, parallel to the line of the fuselage and adjacent to a line of bubble-shaped windows that give searchers a wide view of the forest, tundra or ocean. On the floor in front of the bench is a belly hatch, now closed tightly. Along the opposite wall are racks jammed full of rescue equipment. Sit on the bench, your back to the bubble windows, and watch the crew in the final moments of the horror about to unfold. Because we have slowed down time, their movements will seem slow and inconsequential, but this is the only way to comprehend one of the worst accidents ever to befall the Canadian military's search-and-rescue community. Men who had spent their careers plucking others from disaster are about to be caught themselves at the frontier, with no hope of rescue.

The noise inside the cabin is deafening as the two rotors overhead, each with three blades, claw at the cool October air. The low-flying aircraft — designated Tusker 27 for this flight — has been bouncing around as it digs into headwinds that have exceeded 100 kilometres an hour. This trip began at the airport in Sept-Îles, Quebec, with the pilots heading due south across the St. Lawrence River at a standard cruising speed of about 200 kilometres at hour. Almost forty-five minutes later, as Labrador 305 approaches the opposite shore near the Gaspé community of Marsoui, the winds have become far more onerous. A permanent notice to mariners warns of these localized winds, which can play havoc with navigation and drive vessels onto shore. The aircraft, whose ground speed has been cut to about 130 kilometres an hour as it battles the angry gusts, is being knocked around, making for a rocky ride. The helicopter is flying 150 metres above the ground but will soon need to gain some altitude as it prepares to fly over the Gaspé Peninsula's Monts Chic-Chocs on its way back to base at Greenwood. The fall colours in the forest below are spectacular.

In the cockpit, out of your view, the tip of a T-shaped black handle facing the aircraft commander at his left suddenly glows red, illuminating the word "FIRE." The pilots immediately look for confirmation by watching for smoke emerging from the No. 2 engine, mounted on top of the aircraft on the right hand side. There have been false alarms before, so the direct check is necessary. All on board are alerted through the communications cords that feed the two pilots' voices into their helmet headphones. We may have an engine on fire — No. 2, on the right. The aircraft commander, sitting in the right-hand pilot's seat, makes a slight turn so the crew can see if there's a telltale trail of smoke. And so there is. The snaking black discharge leaves no doubt: the fire is real.

Labrador helicopters are equipped with two engines, not only to deliver enough power for a workhorse aircraft but also to provide a margin of safety in case one engine quits. A single-engine failure is an emergency, of course, but only rarely the cause of a crash. With sufficient forward speed, ideally about 100 kilometres an hour, a Labrador can maintain enough lift with its one remaining engine to manoeuvre to a safe landing site. In a stationary hover, on the other hand, much more engine power is required to stay aloft, and with just one engine, the pilot must get to the ground immediately. Even so, a single functioning engine can get this job accomplished with relative safety. Many helicopters have survived single-engine failures with no injuries or damage, and military training programs routinely prepare pilots for just such an eventuality. Experienced pilots cope with engine failures on numerous occasions throughout their careers. In the case of this flight, conditions are rather good for this sort of emergency: there are several clear landing spots nearby along logging roads, and the aircraft has adequate forward momentum. The two pilots are alert but relatively confident about getting to ground safely and quickly.

With the engine fire confirmed visually, the next job is to extinguish it. The aircraft commander pulls out the T-handle for the No. 2 engine, closing two electrically driven valves in the engine and stopping the flow of oil and fuel. He then twists the T-handle seventy degrees to the left and seventy degrees to the right. Each of these twists triggers separate fire extinguishers that spray the burning engine and surrounding area, snuffing the flames. The T-handle FIRE light goes out, confirming that the blaze is doused and allowing the crew to take the next step, dumping fuel. Canadian search-and-rescue Labradors carry much more equipment than rescue helicopters in other countries and are therefore much heavier in the air. The extra equipment is needed because Canada has relatively few rescue bases spread across a huge expanse, and Labrador crews must fly farther afield, ready for any situation they may encounter. But the drawback is that when an engine quits, the aircraft's weight leaves the pilots very little margin. They must lighten up immediately by dumping fuel.

This procedure is undertaken by the flight engineer, who sits between the two pilots and slightly behind them, with the fuel-management panel in front of his legs. The Labrador's tanks were full with about 2,250 kilograms of fuel when it left Sept-Îles. The forty-five-minute crossing of the St. Lawrence River has burned off some of that, but the aircraft is still too heavy for a controlled emergency landing. The flight engineer calculates that another 450 kilograms need to be jettisoned. "Dump fuel now," the aircraft commander barks into the intercom, and the flight engineer lifts a red safety flap and presses a spring-loaded button on his panel. His calculations tell him he needs to hold the button down for about twenty-five seconds to purge the aircraft of that much fuel. The pilots, meanwhile, keep an eye on the fuel gauges to watch the levels drop.

From the bench, you can see none of this, but the crew chatter spills into your helmet headset, the crisp orders executed flawlessly by a team of professionals. You may be tempted to put your head inside one of the bubble windows to watch the plume of misting grey-white fuel emerge from tubes on the sponsons on each side of the Labrador. The aircraft commander has now turned the helicopter back left toward a valley, completing an S-shaped manoeuvre that began with the check for smoke in the No. 2 engine. He has spotted a clear landing zone, and everyone expects to be safely on the ground in a minute or so. The crew is calm. The emergency landing is an inconvenience, of course, but at least there is a village nearby and roads on which to haul out an aircraft or haul in an engine and replacement fuel. Better here than in the middle of nowhere. One of the search-and-rescue technicians rises from his seat on the left at the front of the passenger cabin and starts to put on his coat to be ready to disembark once they have landed.

This is the point where time must slow down drastically, for in the next few seconds more will happen than any of the crew can possibly comprehend, more than any human observer can hope to witness in the ordinary unfolding of time. The aircraft's simple problem — a manageable engine fire — will escalate profoundly in the blink of an eye, without the least warning. There will be no panic among the crew, no fear, perhaps just a fleeting perplexity and then blackness, a catastrophic but merciful end.

Above you at the rear of the passenger cabin are two rectangular panels in the ceiling that cover the two engines, allowing maintenance crews access to the maze of hoses and tubes on their undersides. The flight engineer can even poke into the compartment in flight, while the engines are running. The stainless steel panels are sealed tightly, allowing no fumes or smoke, for example, to enter the passenger cabin. At the start of this emergency sequence, the right-hand panel suddenly blows from its seating, slightly stretching its fasteners until they no longer hold. The panel smacks the floor of the aircraft, just to your left, but does no real damage. The force of this blast is relatively mild and does no harm to the six crew members, who are all well forward. But the explosion is ominous, coming from an engine compartment supposedly covered in fire-retardant and cut off from fuel or oil. The smoky guts of the idled No. 2 engine are now visible through the ceiling.

Now events happen so fast that no one has a chance even to turn around to see them. The trailing cloud of jet fuel ignites in a massive fireball. The spectacular explosion, its bright orange core embedded in a black cloud, envelops the rear of the Labrador and can be seen clearly from your bubble window. Although it jars the aircraft slightly, the Labrador continues its flight for a few milliseconds, the crew unable to mentally register this second, more powerful blast. The rotor blades, now somewhat underpowered because a single engine drives them, have continued to chop into the cool air, keeping the Labrador steadily aloft. The blade tips travel at about 800 kilometres an hour, each blade making a full revolution twelve times every second. But the blades on the rear rotor have now been caught in this terrible blast. The explosion forces at least two of them upward, out of their disk-shaped plane, as they cross clockwise into the pressure wave of the blast.

One of these long blades chops into the passenger cabin from the left, opposite you and just behind the seat of the standing search-and-rescue technician, who by now has one arm in his jacket. Made of light fibreglass and shaped like a narrow wing, the blade has a metal leading edge along its length. This steel edge, knocked out of alignment by the blast, slices through the shaft at the top of the fuselage that provides synchronous power to the three-blade rotor at the front of the Labrador. It also severs all electrical power from the generators to the cockpit, causing every light and electronic system to fail. Then it cuts into the fuselage like a red-hot knife through butter and carries on out the other side as a second blade begins its own destructive swing through the fuselage, cutting in a slightly different spot. The third rear blade comes around but, not having been hit by the blast like its two companions, misses the fuselage altogether as it remains pretty much in its original plane. The first blade now returns, hitting the search-and-rescue technician and seriously lacerating his arm. The second blade then enters again, chopping the fuselage further. Those parts of Labrador 305 that the blades have not actually sliced now crack apart from the stresses of the flight, and the helicopter "flies" along in two sections separated absurdly by a gap of air. All of the crew members are alive, though one man is injured. Events have happened so quickly, though, that the wound has not had time to bleed. The aircraft looks strange from your seat, like a cracked egg, with the fall colours below now visible where once there was a floor.

The rear rotor blades have lost much of their momentum from slicing into the shaft and fuselage frame, but the front blades, even though they have been denied engine power slightly longer than the rear blades, continue to spin in a counter clockwise direction. Now completely out of synchronization, they themselves are chopped by the wildly out-of-control rear blades. But enough of the front blades are left that one of them comes down on the left side and strikes the external fuel tank on the rear fuselage, cutting it open. A little more than a second or two has elapsed since the engine compartment panel hit the floor, triggering the sequence. The crew remains oblivious.

The shattered left fuel tank now erupts with an enormous explosion, at least five times the strength of the engine compartment blast. And because the Labrador has been chopped in two, its front half containing all of the crew is now fully exposed to this powerful, deadly hammer blow. The shock wave ruptures steel containers and knocks the two pieces of the aircraft farther apart. It also hurls the two search-and-rescue technicians against the bulkhead separating the cockpit from the passenger cabin, killing them instantly. One of these men, sitting in the right-hand chair, had been looking out his bubble window in the direction of the disabled engine at the precise moment of the blast. A flight engineer-in-training, who had been standing over the main flight engineer, is thrown into and over his teacher. Both men die immediately.

The two pilots survive the blast, their backs protected by the bulkhead, but about 30 seconds later it is all over as the front half of the aircraft, its blades trimming the treetops as they continue to spin, hits the ground, bounces sideways about fifteen metres, and hits again outside the main debris field. The rear section, meanwhile, itself breaks into two parts while still airborne. The fin-like pylon on top, with the now-quiet rear rotor blades attached, comes away from the main cabin and the two parts fall to the earth. Unlike the front rotor blades, the rear ones do no damage to the trees on the way down. Because all the fuel was carried in the rear section, a fierce fire breaks out on the ground. The magnesium case of the transmission ignites, burning hot and bright.

▪ ▪ ▪

This, then, was the view from the aircraft itself. The view from the ground was almost as dramatic and horrifying.

Labrador 305's mission began as a medical evacuation flight to pick up a female patient from the remote north-shore Quebec community of La Romaine, which is connected to the rest of the province only by ferry. The crew left the Greenwood air base shortly after midnight on October 2, 1998, refuelling at the town of Gaspé, Quebec. On attempting to cross Anticosti Island in the Gulf of St. Lawrence, the pilots ran into such strong wind that they landed at Sept-Îles to await better weather. The crew finally made it to La Romaine at nine-thirty a.m. where they brought aboard the patient on a stretcher. A male family member and a nurse accompanied her for the bumpy trip. Winds continued fairly strong, making the nurse ill and forcing the Labrador to make another fuel stop at the small airport near Natashquan, Quebec, before carrying on to the hospital at Sept-Iles. An ambulance greeted the crew on the tarmac and the three La Romaine passengers were quickly taken to the local medical centre. Despite the buffeting, the Labrador ran smoothly for the entire flight. "There weren't any problems," said Sergeant Tim Eagle, one of the two search-and-rescue technicians on board. "It ran like a top."

The five-man Labrador crew had by now been on duty for more than thirteen hours through the night and had to be relieved, so a Hercules aircraft from Greenwood flew to Sept-Iles with a fresh crew. The medevac crew, who had eaten only box lunches while flying the

Labrador, now got a hot meal and soon transferred their personal equipment onto the Hercules for the return flight to their Annapolis Valley base. Eagle was by now exhausted, and after the seatbelt lights went off he settled into a deep sleep in a bunk at the rear of the cockpit of the droning Hercules. Never before had he been invited to sleep up front in the officers' space, and he relished the opportunity to stretch out and doze. Eagle remembered groggily watching the array of dials in front of the pilots for a few minutes, thinking with some satisfaction that he had finally gotten through an arduous night and day. "Wow, the whole thing's done," he thought before drifting off, overcome with fatigue.

Labrador 305, meanwhile, lagged far behind the Hercules because of its slower maximum cruise speed. The Hercules crew had offered to fly more slowly to accompany the Labrador, but the helicopter crew radioed that all was well and the Hercules should carry on ahead. The Labrador commander piloting from the right-hand seat was Captain Darrin Vandenbilche, thirty-three, from Invermere, British Columbia, who was engaged to be married. On the left was co-pilot Captain Peter Musselman, also thirty-three, from Edmonton, married with one child. Master Corporal David Gaetz, thirty-seven, was the flight engineer, responsible for all of the aircraft systems as well as for rescue hoisting. Gaetz was married with three children. Working alongside him was Master Corporal Glen Sinclair, thirty-six, who was in training as a flight engineer, making a sixth crew member aboard the Labrador, which normally carries only five. Sinclair was also married with three children. Lastly, there were the two search-and-rescue technicians: Master Corporal Darrell Cronin, thirty-two, from St. John's, married with two children; and the senior Sartech, known as a team leader, Sergeant Jean Roy, thirty-four, from Montreal, with a wife and three children. Cronin was seated in the cramped left-hand seat behind the radio equipment rack in the back of the Labrador. Roy was in the team leader's seat, with more leg room, on the right. All of these men knew each other well, having worked many training and operational missions together.

The rescue squadron at Greenwood does many fewer medevacs these days than even five years ago, mostly because provinces including Nova Scotia and New Brunswick have finally accepted the responsibility of providing their own air-ambulance services. For years, the cheaper alternative was to leave it to the military rescue squadrons, which often had to put premature babies and heart patients inside the rumbling, chilly cargo hold of a Hercules transport. The Canadian Forces would always bill the provinces for such emergency services, but few provinces ever paid. It had been a polite game played between two levels of government, with the military always stuck with the bill. Some remote evacuations will always be the responsibility of the rescue squadrons, but the overall numbers of medevacs have declined dramatically as provinces have shouldered more of their share of the emergency health burden. Labrador medevacs were therefore becoming a far less regular part of the Greenwood squadron's routine.

Residents of the small Gaspé community of Marsoui, Quebec, rarely saw rescue Labradors passing overhead — even fewer now than in the past — so it was a noteworthy event to see one approaching the village that crisp Friday afternoon in October. At least thirty-two people, some as far away as five kilometres, watched as the chopper with the distinctive yellow and red markings completed its transit of the St. Lawrence, crossed Marsoui on its way over the Chic-Chocs. Winds were strong, gusting to more than 100 kilometres an hour, slowing the aircraft markedly. In addition, the Labrador was seen being jostled by mechanical turbulence, the high-energy eddying and shifting of blowing air as it is interrupted by mountains, buildings or other ground obstacles. The aircraft continued its southward track more than five kilometres beyond the village toward the Marsoui River valley, where at least seven witnesses continued to follow its journey. Suddenly, most of them noticed a puff of smoke out the rear and watched as the aircraft started a turn. Then a massive explosion or fireball engulfed the back of the Labrador and the machine came apart in the air immediately afterward. Three separate pieces fell quickly and almost vertically into the dense forest south of Marsoui. "I saw the helicopter pass over my house; there weren't any problems," said Alberto Henley, a worker at the local sawmill. "It wasn't going fast, but it was going. Then it went down and there was an explosion." Another sawmill worker, foreman Réjean Leclerc, was among the first to reach the bodies. "It's not something I'd like to see often," he said that night.

Wreckage from the disaster was strewn over an area a kilometre long and about half a kilometre wide along an uphill path that rose about 200 metres. Each of the three main pieces of the aircraft fell across a 185-metre stretch, with the aft section of the fuselage lowest down the heavily forested slope. Here a fierce fire raged. Forty-seven metres farther uphill was the aft pylon with rotor blades, and another 138 metres higher up was the cockpit section with the limp bodies of all six crew. A minute after the crash, at least one of the witnesses called the Quebec provincial police in nearby Ste-Anne-des-Monts, who notified the military's

Canadian Mission Control Centre in Trenton, Ontario. About five minutes after the crash the centre called the Halifax rescue co-ordination centre to report that there had been an accident with an unidentified helicopter. Fears were immediately aroused that this might be Labrador 305 when witnesses referred to the yellow fuselage. Halifax called the coast guard rescue centre near Quebec City to see whether there might have been any coast guard choppers in the area, but the nearest was far to the east over Anticosti Island.

If this were Labrador 305, where was the rescue beacon? Each military aircraft is equipped with a device called a Crash Position Indicator or CPI. The instrument is designed to jettison in a crash and emit a continuous signal on the 243 megahertz frequency that can be picked up by the low-orbit SARSAT satellites. The system also detects a civilian aircraft version of the rescue beacon using Emergency Locator Transmitters broadcasting on the 121.5 megahertz frequency. In addition, these transmissions — both military and civilian — can be picked up and homed in on locally, though the first alert is intended to be by satellite. In the Marsoui crash, the satellite system seemed not to pick up any emergency beacon — raising hopes that perhaps the felled aircraft was not the Labrador after all.

Sergeant Tim Eagle was awakened in the Hercules when it was just twenty minutes away from the Greenwood base to be told they were heading back toward Sept-Îles. The Halifax rescue co-ordination centre had called several times, first to ask whether the Hercules crew was in contact with Labrador 305. The pilots then tried to radio the helicopter crew, but there was no response, not really surprising considering the distance of hundreds of kilometres. Another call from Halifax said there was a report of a helicopter crash near Marsoui. The Hercules crew was asked to turn back toward Sept-Îles "hastefully." Master Corporal Darcy St-Laurent, a search-and-rescue technician on the Hercules, remembers feeling excited about the prospect of a mission, without any inkling that the helicopter might be their own. In yet another call, the rescue co-ordination centre passed on the information that witnesses had said the helicopter that crashed was red and yellow, a detail that stoked a nagging fear among the Hercules crew members.

"At that point, we had a suspicion that it was our company aircraft," recalled Sergeant Maurice Robert, another search-and-rescue technician aboard the Hercules. "It's a very uncomfortable feeling, a very worrisome feeling." Eagle remembers the pilots using the Hercules's public address-system, an unusual procedure, to provide the few details as they were known. "I could hear something in their voices," he said. "But I never would have thought in my wildest dreams what we were going to see." The Hercules pilots put the aircraft on maximum cruise as Halifax acknowledged that the downed aircraft might be Labrador 305 by saying it was "possibly company traffic." As the Hercules crew approached the crash zone about forty agonizing minutes later, their headsets picked up the continuous woo-woo-woo-woo signal of a military rescue beacon. It had become clear that the Labrador was down, but what of the crew? Could they be saved?

Near Marsoui, the Hercules came in low to try to locate the crash site, buffeted heavily by winds, which reached seventy-five kilometres an hour. The cockpit crew could see a smouldering fire on the ground below. Flames became visible inside a blackened circle 10 metres in diameter carved into the thick forest floor. Robert and his Sartech partner, St-Laurent, now prepared to jump into the scene by parachute, a tricky manoeuvre in the high winds. The Hercules had to remove itself from the mechanical turbulence close to the ground, so the crew decided to make the jump from about 800 metres, where the winds were stronger but less fickle. Even so, while the Sartech manual permits training jumps only where winds are less than about fifty kilometres an hour, for safety, the winds for this operational jump would be up to twice that threshold. The pilots brought the aircraft over the crash site, and the Hercules's loadmaster jettisoned some streamers and a C-8 cartridge, which deploys a three-metre dome parachute carrying a smoke-emitting canister. The C-8 produces a clearly visible, nine-minute-long smoke trail which does not create fires on the ground. Maintaining a constant speed, the Hercules pilots then began a run directly into the wind from the point at which streamers and C-8 hit the ground back to the target, with Eagle counting the seconds. Once they were over the target, Eagle counted the same number of seconds as the Hercules carried on at a steady speed, bringing the jumpers to the point where they knew the wind will carry them back to the target. The wind was so strong that day that the two Sartechs would have to drift about five kilometres back to the crash site.

Robert and St-Laurent stood on the lowered ramp at the back of the Hercules, the interior of the aircraft exposed to the chilly outside air. This would be only the third operational jump for Robert, only the second for St-Laurent. A thousand and one, a thousand and two, a thousand and three — Eagle had to count off thirty-five seconds for the aircraft to fly upwind those five kilometres. The two anxious men standing on the

Sartech Master Corporal Darcy St-Laurent checks his gear aboard a Hercules at Greenwood, Nova Scotia. DB

ramp awaiting the signal were incredulous that the set-up was taking so long. "I remember them standing on the ramp saying, 'What's taking so long? We should go, we should go!'" Eagle says. The impatient pair wondered whether the jump had somehow been cancelled. Finally, at thirty-five seconds, Eagle gave a hand signal and the two men jumped, first St-Laurent, then Robert, their rectangular chutes automatically deploying seconds after leaving the aircraft because of a static line to which they had attached their ripcords. "It was a hairy jump because it was so windy," Robert recalled.

They were so far away from the target that it was difficult to become oriented as they carried out their safety drills in the air. There was simply no visible smoke to aim for. "I was looking at the wrong mountain ridge," said St-Laurent, who guided his chute off course until he noticed the white trail of smoke from the crash. He corrected his flight path, but as he came closer to the ground the turbulence became fierce, causing him to oscillate under his canopy. He landed slightly downwind of the crash site on a logging road, where he was protected from the wind, and radioed the Hercules to report his safe arrival. Nearby, there was a cluster of emergency vehicles that had arrived about thirty-five minutes after the crash and about twenty-five minutes before St-Laurent appeared. The paramedics and police directed him to the wreckage. He first came across the aft pylon with rotor blades attached, then found a mass of burning wreckage. The blaze had consumed so much of the aircraft that St-Laurent could identify only a few pieces, and he immediately thought that the charred, twisted metal represented the whole of the aircraft. There were no bodies, however, raising hopes that the crew had somehow gotten away unharmed. A few minutes later, St-Laurent organized a ground search team to sweep up the mountain toward where some people had been told there were bodies. He kept calling out the names of the missing crew members, his colleagues.

St-Laurent's partner Robert, meanwhile, came down closer to the crash site, slightly downhill. His parachute got tangled in a tall maple tree, leaving him dangling about eight metres in the air. His rugged bush suit had an Elizabethan-style high collar that protected his neck and head, so he was caught but uninjured. Robert reached into a pocket and pulled out a Sky Genie rappelling kit, an eighty-three-metre coil of rope and with a clip originally designed to let window-washers rescue themselves in an emergency. Standing firmly on a branch, Robert quickly extracted himself from his harness and then from the tree. Unfortunately, he left his hand-held radio behind in a pocket on the harness, leaving him unable to contact either his partner or the worried Hercules crew above. The aircraft, meanwhile, kept circling the area looking for a safe drop zone in which to parachute in medical and survival equipment in case it was needed. The crew knew the Labrador wreckage was on fire but hoped against hope that the blaze had broken out on the ground after everyone had gotten out safely.

Once Robert was on the ground, he had walked only about two minutes downhill looking for wreckage when he met a civilian who had witnessed the disaster. The man took him directly to the cockpit section with all six bodies, and Robert frantically began to look for signs of life among the limp forms of his colleagues and friends. The bodies were still warm. "I really stayed long with each person to make sure I had no other signs that might indicate life," he said. "I knew them all . . . they were all readily identifiable." None of the six showed the slightest tremor of life, as Robert cut open the clothing at their chests to apply a stethoscope. "It appeared that they all died of sudden deceleration," he said later. "They also sustained injuries from a high impact." In the forty minutes he was alone among the bodies, Robert checked and rechecked everybody, looking in vain for a feeble pulse or a dilating pupil. These were the worst moments in the most traumatic search-and-rescue mission that Robert ever had to undertake.

St-Laurent by this time had found his partner and learned the grim news. To be absolutely sure, he did his own check of each body, placing his own stethoscope on each man's chest, examining the pupils, checking for breathing. Some of the men were badly mangled, but others looked as if they had just fallen asleep. "It's a very difficult task to do because you know all these faces," St-Laurent said later. He gave Robert the hand-held radio so his team leader could let the anxious Hercules crew know the extent of the disaster. This was also a military requirement, to file a Notice of Crash Location or NOCL, confirming the identity of the aircraft and the status of the crew and passengers — "six black." "It did bother me when I made the call to the Herc," Robert said. "That call was very hard for me to do." Inside the Hercules, an hour of taut hoping collapsed when the radio crackled with the grim news that every crew member was dead. "That message, that was very sad," said Eagle. "You could really hear that in his voice. He was holding back the tears — we could tell he was close to tears." The message, that six close colleagues were gone forever, ripped through the aircraft. "It was devastating," says Eagle. "I couldn't believe what I was hearing. It's a horrible thing. It goes right to the pit of your stomach."

With dark setting in and the Hercules running low on fuel, the crew now headed for Sept-Îles. Robert and St-Laurent later got a lift to Sept-Îles aboard a small Griffon military helicopter from Bagotville, Quebec, brought in to help secure the crash site for evidence. The winds continued unabated for the trip across the St. Lawrence, badly jostling the two Sartechs. "I just don't like helicopters any more," thought St-Laurent as he tried to make sense of the disaster. In Sept-Îles, pizza was ordered and Robert and St-Laurent told their colleagues about what they had seen — everyone, that is, except the Hercules cockpit crew who still had to get the aircraft back to Greenwood that night and did not want morbid details to impair their focus. "You're numb," said Eagle. As the Hercules passed over the Marsoui crash site, its radio again picked up the woo-woo-woo-woo of the Labrador's still-wailing rescue beacon.

A Sea King helicopter at the Shearwater air base — which was already on standby for the operation to recover Swissair Flight 111 wreckage — was also put on standby for search and rescue, since the grieving 413 Squadron crews would not be asked to do any missions for some time. Back at the Greenwood base, counsellors were made available for crews and families. All of the fifteen remaining Sartechs at the squadron gathered to greet the four Sartechs getting off the Hercules, and they talked until the early morning. "It was nice to see all the boys there," said Robert. Debriefings continued through the weekend to try to console the shattered spirits of the men and their wives. Search-and-rescue crews were brought in from the Trenton air base — no one at Greenwood would be able to undertake any Labrador missions for the next ten weeks. The squadron was already in mourning for a colleague who had died that summer. Don Richard McKinnon, a forty-four-year-old warrant officer, was killed in a July traffic incident while in Guyana delivering humanitarian aid, and several other squadron members with him were badly injured. The enormity of the Swissair crash in early September, which had drawn most of the squadron to the shores of St. Margaret's Bay only to find body parts, had deepened the sense of gloom. The squadron seemed jinxed, leaving wives and families nerve-wracked and fearful of the future.

■ ■ ■

Major Jim Armour had a royal mess on his hands. Appointed the air force's chief investigator for the crash of Labrador 305, Armour faced a logistical nightmare: thousands of pieces of wreckage strewn a kilometre or more across a remote, thickly forested mountainside. Any single piece of wreckage might hold the vital clue to the disaster, so all had to be found, identified, plotted, photographed and recovered. To add to the confusion, most of the rear half of the fuselage had been consumed by a fire that was contained but still smouldered days later. The army had already secured the site, and some flight-safety officers were interviewing witnesses when Armour and his team flew to Mont Joli, Quebec, from Ottawa, arriving at the Marsoui site about eighteen hours after the crash. With him were Major François Tremblay, chief technical investigator, and Major Brian MacDonald, a former Labrador pilot himself who was in charge of operations. The number of people put into the field in the week following grew to more than seventy-five, including up to eighteen investigators. But in the end it would be up to Armour and his two officers to sort out the puzzle of Labrador 305.

A makeshift headquarters was established in the offices of the local Serie GDS sawmill, where some witnesses made statements, but the team later moved to the Marsoui community centre. A tent was set up near the crash site to serve as a field command post, and a kilometre of road was cut into the bush so the wreckage could be extricated. For five days, four teams of five people each moved through the tangled bush, mapping the wreckage using Global Positioning System receivers for precision. The ejected rescue beacon was still emitting its signal days after the crash, lodged under some wreckage that could not be disturbed for fear of erasing clues. The Halifax rescue co-ordination centre, meanwhile, had determined that a Russian rescue satellite was passing overhead at the time of the crash. The satellite, dating from the days of the Cold War, was purposely designed not to detect Western military rescue frequencies, only the civilian frequency. Thus this orbiting relic of the superpower struggle was simply not equipped to pick up the beacon of the Labrador and alert Halifax rescue officials to the crash of their own rescue chopper.

The bodies had all been taken to nearby Ste-Anne-des-Monts for a preliminary medical examination, then on to Montreal for detailed autopsies, presided over by flight surgeons, to extract as much information as possible about their final moments. The autopsies were far more meticulous than those conducted after routine civilian deaths. Bodies in an air disaster act like biological black boxes. The injuries mutely record many details of the last seconds of a flight and are as vital to an investigation as the final reading of a gauge or the crash position of a cockpit lever. Military police flew in from Greenwood to gather the personal effects of the crew and to retrieve the rifles and morphine from the rescue kits of the two search-and-rescue technicians. A roster of more than thirty witnesses was drawn up, and Armour's team began to try to piece together the fragmented evidence. In the military's small search-and-rescue community, it was the worst disaster since June 14, 1986, when eight people perished after a Twin Otter from 418 Squadron crashed west of Calgary while searching for two missing aircraft in the Rocky Mountain foothills. Five of the dead were civilian volunteers.

The pressure on Armour's group was enormous. The crash inevitably migrated into the political arena, since the aging Labrador and Sea King helicopters had all been due for replacement by forty-three high-tech EH-101s — but the $4.8-billion order had been cancelled in 1993 by the newly minted Liberal government, in fulfillment of a campaign promise. The search-and-rescue and Sea King squadrons were thus left in the lurch. During the 1993 election campaign, Liberal Leader Jean Chrétien had railed against the Conservative decision to buy the new helicopters, saying Canada could not afford them. But then his government dragged its feet for more than four years before finding its own replacements, finally placing an order in April 1998 for fifteen search-and-rescue helicopters, the first to be delivered in 2001. The Liberals' choice for a new search-and-rescue helicopter was a stripped-down version of the same three-engine EH-101, now called the Cormorant. (There was no immediate replacement for the Sea Kings, which are the same vintage as the Labradors but arguably in worse shape.) Had the original EH-101 order been allowed to stand, some of the geriatric Labradors might have been retired by the time of the crash. Military families, especially relatives of the dead airmen, joined opposition parties in fixing the blame on the Liberals for playing politics with men's lives. And much of the purported cost saving on the Cormorant deal was in fact eaten up by $500 million in EH-101 cancellation fees and the rising maintenance costs of keeping the Sea Kings and Labradors in the air.

The political pressure was raised a notch with reports that the senior flight engineer on Labrador 305 had been expecting a disaster. Master Corporal David Gaetz apparently kept a private journal of Labrador mechanical problems, at least according to his father-in-law. "He told me it was secret," Gerald Daley of Wilmot, Prince Edward Island, said in an interview the day after the crash. "He didn't want anybody to know because he tried so hard to get changes and do things for the Labs. He knew they were in bad shape and everything. And talking to the big brass and so forth, I guess they just didn't listen." Daley said Gaetz made plans about where his computer diary would go in the event of his death, and that Gaetz's wife Roslyn did not control the file. Gaetz was "scared that certain ones in high places want to scratch it out or strip it," Daley said. "This is why he wanted to keep it secret." Daley's claims were later denied by Gaetz's wife Roslyn, who called it a "media fabrication." But documents released under the Access to Information Act four months after the crash suggest that Gaetz had made at least some of his concerns known to his superiors. He was particularly concerned about a "power check" procedure, a ground-based method of deter-mining whether the engines were safe to fly before takeoff. The air force, in fact, had made changes to the procedure when it upgraded all of its Labrador engines. The concerns Gaetz had voiced to his bosses thus appear to have been resolved before the crash of Labrador 305. In any case, the so-called secret diary — which presumably had not been revealed to Gaetz's commanding officers — has never surfaced, and Gaetz's father-in-law stopped talking to reporters about it.

Maintenance records for Labrador 305 showed problems that were routine and relatively benign. The most recent incident was a September 29, 1998, precautionary landing at Yarmouth, Nova Scotia, because of a flight-control concern. Another incident on May 22, 1997, showed the T-handle fire light illuminating twice — falsely indicating an engine fire during flight. Generally, the maintenance file was of little guidance to investigators, though it did demonstrate the myriad problems that can crop up unexpectedly with any older aircraft. A Labrador flight engineer at Greenwood, who preferred to remain anonymous, put it this way the day after the crash: "I think that with any kind of a machine that takes abuse for a number of years, there are definitely going to be hidden factors that nobody can foresee."

The fleet of Labradors, reduced to twelve after the Marsoui crash, had indeed been showing its age. The first six had been ordered for the Royal Canadian Air Force in 1960, primarily for search and rescue, and deliveries were made between October 1963 and November 1967. The Canadian army ordered a slightly different model, then known as the Voyageur, in 1963 and took delivery of twelve in 1964 and 1965. Built by the Boeing Airplane Co. of Morton, Pennsylvania, the twenty-five-metre-long aircraft used new technology for its time. Its twin gas-turbine engines were far more reliable than traditional piston engines, for example. But by 1998, the Labradors' 1950s-era technology had become positively archaic. Although about 450 of these aircraft were still flown by the military in the United States, Sweden, Japan and Saudi Arabia, as well as by an American commercial outfit, these operators had regularly updated the avionics and other technology.

By the mid-1990s, the 1,350-horsepower engines on the Canadian versions provided too little power, given the amount of rescue gear aboard, and the engine model was used nowhere else in the world. Parts had become almost impossible to procure, and maintenance crews sometimes had to resort to machining their own parts from scratch. The aircraft needed an average of eleven hours maintenance for every hour in the air. The cash-strapped Canadian air force had been cancelling technological updates on the Labradors since the late 1980s, assured that a replacement helicopter was on the horizon. "A lot of decisions were made back then to not procure anything because something was right on the heels of the Labrador," said one senior maintenance officer. When the EH-101 program was abruptly cancelled in the fall of 1993, the military was stuck with an aircraft long overdue for upgrades and bristling with orphan technology.

Canadian military officials often stated that some operators of Labradors in other countries intend to fly them until 2020, and that some have been flown as many as 56,000 hours, whereas the Canadian average at the time of the crash was about 15,500 hours. (Labrador 305 had flown slightly longer, 16,434 hours.) But apart from ignoring the fact that other countries had been upgrading their technology continuously, the claim did not take into account Canada's specialized use of these aircraft. No other country regularly punishes their Labradors with search-and-rescue work, which, by its very nature, places them in some of the worst weather imaginable. The hostile environment includes salt spray, one of the most corrosive assaults that an airframe can endure. The Greenwood and Gander Labradors, which undertake far more marine rescues than the other Labrador squadrons in Trenton and Comox, are especially prone to salt corrosion.

Known as the "rust belt" in the military, the East Coast environment was especially dangerous for an older helicopter. "Corrosion damage generally increases exponentially with time, and as an aircraft becomes older the effects of corrosion become more severe," an April 1999 military assessment of the Labradors acknowledged. "A particularly serious consequence of corrosion is that it can accelerate other forms of damage, such as [metal] fatigue, and acts conjointly with fatigue to lower the overall structural integrity of the aircraft. . . . These factors in isolation or combination can result in catastrophic structural failure." Canada's military maintenance technicians, among the best in the world, thus face a constant battle to obtain parts and stay one step ahead of problems. Despite their skill and diligence, Labradors are available only about sixty per cent of the time because of persistent breakdowns.

Indeed, the aircraft had been involved in two other high-profile crashes in the 1990s that underscored its obsolescence. On April 30, 1992, a Labrador lost power in one engine while on a rescue mission in the mountains about twenty-eight kilometres southeast of Bella Coola, British Columbia. Caught in a stationary hover, the pilot had no choice but to get to ground immediately. A Sartech, Corporal Phil Young, was thrown from the aircraft, struck by a rotor blade and killed as the Labrador hit the ground and rolled up to six times. The aircraft ended up more than 100 metres downhill, upside down in a deep snow bank. Two passengers inside the aircraft were seriously injured, and a fire broke out on the ground, fed by a ruptured fuel tank. A lengthy investigation — hampered by the isolated location — failed to determine a precise cause for the engine failure, and the accident today remains a mystery. The investigators suspected an engine-related problem but could not nail down the source.

And on May 1, 1995, another Labrador plummeted eighteen metres into trees at Margaretsville, Nova Scotia, on the Minas Basin, while in a hover on a routine training flight. No one was seriously injured, thanks in part to the pilot's quick action in moving the helicopter forward, thus avoiding the two Sartechs standing on the ground below. An investigation later found that an engine failed because a loose clip had been sucked into the turbine, the fault of a private contractor who had recently overhauled it. But the investigators also found that the aircraft had experienced six engine-related safety problems in the previous fourteen months. In addition, the inquiry uncovered the discomforting information that the fleet of thirteen aircraft had experienced forty-six such problems over the previous two years, a worrisome trend.

Yet another Labrador was forced to land in a field north of Kingston, Ontario, when one of its engines quit on November 1, 1995. No one was hurt, but these three incidents so spooked the search-and-rescue squadrons that air force brass finally set up an eight-man investigative team to find out why the engines kept sputtering out. The team's meticulous report in April the following year confirmed what everyone already knew: power losses were happening too often on the Labradors, and crews could expect major in-flight hovering problems, such as that which precipitated the Bella Coola crash, about once every two years. The number of engine-related safety incidents began to rise significantly in 1991, the team noted, about the time the cost-conscious air force scaled back on engine upgrades, knowing that a replacement chopper fleet was just around the corner.

Indeed, the report cited a decision by Air Command in early 1992 to forgo engine upgrades because of the imminent delivery of EH-101s. Other operators around the globe had been more vigilant. "It is evident that the CF [Canadian Forces] is the only operator in the Western Hemisphere not to have initiated an extensive engine control system modification of some sort," the investigators concluded. "Although the aircraft is soon due for replacement, it would be unacceptable to maintain the status quo as this [engine control] system has demonstrated an uncommonly high degree of unreliability. The technology and design of the current system dates back over forty years and no longer exhibits the necessary degree of reliability." A few months after the damning report, military brass finally agreed to improve the engines at $200,000 a pop, though the resultant increase in horsepower to 1,500 would still not allow for a single-engine stationary hover. Four of the thirteen aircraft had yet to receive the new engines by the time of Labrador 305's crash. Labrador 305 itself had not yet been fitted with the new, more reliable power plants.

The troublesome General Electric engines were also used on the Sea Kings, the ship-based submarine hunting helicopters that were virtually the same vintage as the Labradors. In 1994, months after the Liberals had cancelled the EH-101 order, a single-rotor Sea King crashed near Saint John killing two crew members after a particularly harrowing emergency landing. An investigation later determined that a fuel line had developed a leak, causing an uncontrolled engine fire during flight — a problem similar to Labrador 305's apparent trouble. Smoke from the fire had built up in the cockpit, blinding the pilots, who were forced to open a window — which in turn fed more oxygen into the blaze, turning the aircraft into a flying blowtorch. Eventually new, more robust fuel lines were designed to prevent a recurrence, but the episode demonstrated how unpredictable an aging aircraft could be. Technology updates for the Sea Kings had also been postponed since about 1989 and for the same reason. The Liberal chopper cancellation suddenly caught the military with two fleets of hopelessly obsolete helicopters and with no upgrades in sight.

These issues weighed heavily on Armour and his team. The search-and-rescue squadrons desperately wanted to know whether their Labradors were safe to fly. The entire fleet had been effectively grounded hours after the crash in the hope the investigators could come up with some quick, decisive analysis. Labradors could be launched only in search-and-rescue operations where life was at risk and the aircraft was "absolutely required." Military brass also urgently needed to supply answers to their nervous political masters. To add to these pressures, Armour could not rely on cockpit voice recorders or flight data recorders to help resolve the enigma. This technology was not offered at the time of the original purchase, and the air force decided it was too expensive to retrofit its Labrador fleet with these "black boxes." In addition, Armour could not interview any survivors, who are often the best source of information about the final moments of a troubled flight. Rather, there were only civilian witnesses on the ground, who had no specialized background in aviation, and some of whom were as far as five kilometres away from the crash site. There was no radar information — the aircraft was flying too low to register — and there were no radio communications in the final minutes of the flight. The investigation was almost entirely reliant on the shattered debris spread across a thick mountain forest and on the patterns of injuries on the dead.

At the outset, Armour's team could be sure about only four things in this disaster. First, there was an engine fire. From the position of the throttle in the wreckage, it was clear that the pilot had shut down the No. 2 engine on the right hand side of the aircraft. The two fire bottles had also clearly been emptied, spraying extinguisher over the engine compartment. Second, there was some kind of explosion that could be seen externally. Most of the witnesses were clear on that point, some of them reporting a huge fireball. Third, rotor blades had hit the mid-section of the fuselage at some point. Marks on the wreckage made this almost a certainty. And fourth, the aircraft had broken up in the air and fallen to the ground in pieces. All of this happened within seconds, the witnesses said. The difficulty was to stitch these events into a probable sequence that accounted for all of the evidence and explained why each step happened. Perhaps the biggest challenge of all was to account for that massive explosion, for which there was no obvious source.

Four days after the crash, one of the ground teams found and catalogued two fuel-dump tubes, each about the size of a forearm. The tubes are normally located in the rear of the sponsons, the bomb-shaped projections on each side of the Labrador fuselage that contain the rear wheels and external fuel tanks. The tubes normally sit inside the sponsons but can be extended outward during fuel-dumping to help keep the volatile fluid away from the aircraft body as it trails behind. Once extended for fuel-dumping, the tubes cannot be retracted again from the flight engineer's panel, and thus their position provides an important clue in a crash. Both tubes in the Labrador 305 wreckage were

A sponson on a Labrador showing the fuel-dump tube. DB

recovered in their non-extended position, that is, they had apparently remained inside the sponsons — an indication that the crew had not been dumping fuel at the time of the disaster. Perhaps events had happened so quickly, the investigators reasoned, that there had been no time to get to this part of the single-engine emergency checklist. Perhaps the aircrew was coping with single-engine flight in another manner. It was curious, but only one piece of information in an enormous, jumbled assortment of scattered debris.

After two weeks of on-site wreckage recovery, surveying and examination — often in rain and wet snow — it became clear to Armour's team that there would be no quick or easy answers. The engine had clearly given trouble, but there was no investigatory path that took them from a quite manageable engine fire to an explosion powerful enough to knock the aircraft out of the sky. The team would need more time, a more controlled environment and laboratory facilities to move the inquiry ahead. Armour and his boss, Colonel Michel Legault, then head of the air force's flight safety directorate, decided on a highly unusual procedure:

every piece of wreckage would be hauled by truck to Ottawa, where technicians would re-assemble the Labrador piece by piece to try to glean more clues. "We've looked and we've gone many a time over the wreckage site and haven't found the key to this mystery," Legault said soon after the decision was made to rebuild. "There are no other clues to help us. It's a bad one. Very tough." The rebuild operation began October 17, as a flatbed truck and army vehicles began the long journey from the Gaspé to Hangar No. 5 at Ottawa's Uplands military airport. The hangar, which the air force had been in the process of abandoning, was brought back to life again five days before its scheduled decommissioning; reconnecting electricity and heating alone cost about $90,000. The engines, cockpit instruments and other key parts, meanwhile, were taken to a military laboratory in Hull, Quebec, for detailed inspections. Armour's "tin kickers," as crash investigators are affectionately known in the air force, were being given every resource to find the cause of Labrador 305's demise.

Meanwhile, the air force lifted most of its Labrador flying restrictions on October 27, more than three weeks after the crash, even though the cause of the disaster remained unknown. The commander of the air force, Lieutenant-General David Kinsman, said the choppers would be restricted to rescue duties and the training essential to keeping crews qualified. "There is no doubt that the future of the Labrador has been one of the most difficult decisions during my tenure as chief of the air staff," Kinsman wrote in a message conveying the order. But there would be no flying immediately for the shell-shocked Labrador crews at 413 Squadron in Greenwood who were given special leave because of the disaster. Sea King helicopters based at Shearwater would shoulder the load for more than two months, even though their own availability rate was a pathetic forty per cent because of persistent mechanical problems. The Sea King also did not normally carry medically trained search-and-rescue technicians among their four-person crews and had only the most basic rescue equipment on board. (However, Sartechs and extra medical equipment were placed aboard a Sea King stationed temporarily in Yarmouth for the November start of the southwestern Nova Scotia lobster season, a job normally handled by a Labrador.) Air force officials had asked the Agusta-Westland factory in Vergiate, Italy, which was building the new Cormorants, whether delivery could be speeded up. But they were told that the 2001 scheduled delivery of the first aircraft was already the earliest realistic date. Several private companies offered to lease interim search-and-rescue choppers to the military, but the costs and training requirements were deemed prohibitive. Instead, the Sea Kings and the fleet of ninety-nine light-duty Griffon helicopters would have to fill any rescue gap.

At the same time, the air force ordered a fleet-wide inspection of its Labradors. Maintenance crews were required to conduct a nose-to-tail examination of numerous systems that might be related to the crash of Labrador 305 and to have zero tolerance for any deviation from established standards. The inspection was to be carried out as soon as possible on the Labradors that were not on search-and-rescue standby and on the remaining aircraft as soon as possible after they were removed from standby status. This so-called omnibus inspection turned up many problems, especially with engine compartment wiring — so many, in fact, that a second fleet-wide inspection was ordered on the engines to ensure that all wiring and related hardware was compatible with the high temperatures of the turbine power plants. The very first aircraft inspected, Labrador 304, was found to have "numerous non-compliant wires." The stepped-up inspections were an additional burden on air force technicians already pressed because of parts shortages and aged airframes.

Senior maintenance officers for the Labrador, meanwhile, ordered an urgent study of the reliability of the airframe in the wake of the Marsoui crash. Three specialists were to examine ten years' worth of maintenance records across the fleet and determine statistically what Labrador systems were giving problems. The study also included 902 flight safety incidents reported over the decade. Their report, delivered November 19, showed that the archaic engines were the "least reliable item" on the aircraft. But the document also concluded that since the introduction of engine upgrades in early 1997, fleet problems had declined markedly. The results had a calming effect at the top echelons of the air force. Labrador 305 had been awaiting an engine upgrade, but at least the rest of the fleet would soon be fitted and would presumably be in good shape to carry on safely with search and rescue.

The Greenwood squadron's commanding officer, Lieutenant-Colonel Mike Dorey, first test-flew a long-idled Greenwood Labrador on November 13. But mechanical problems continued to plague the squadron's choppers, including a fuel leak, a minor fire and broken screws. The squadron finally resumed search-and-rescue duties on December 16 — more than ten weeks after the Marsoui crash — when a Labrador crew was assigned to find the wreck of a Cessna 172 in dense woods near Liverpool, Nova Scotia. Tragically, the two brothers aboard were found dead. "I thought we'd get a few day missions first to warm up a little bit,"

General Maurice Baril, left, chief of defence staff, and Major Jim Armour examining the wreckage of Labrador 305. DND

said the primary pilot, Captain Doug Parker. "We were shoved right into a very difficult mission at night for our first one. And it went very well. There was no hesitation at all." The choice of Parker as aircraft commander for this comeback mission was particularly apt: he had flown Labrador 305 for the La Romaine medevac before being relieved in Sept-Îles by the doomed crew. With the Labrador fleet reduced to eleven operational airframes — one Labrador at a time would normally be in for a complete overhaul at the Boeing facility in Arnprior, Ontario — maintenance crews at the four rescue bases were now under intense pressure to keep the helicopters airworthy. Kinsman, meanwhile, ordered an accelerated engine-upgrade program to ensure the final few Labradors got the new power plants by February 1999.

■ ■ ■

The reconstruction of Labrador 305 on a makeshift jig in the Uplands hangar had been carried out with painstaking exactitude over the better part of a month. The effect of the ground fire at Marsoui was immediately clear even to a layman — almost the entire rear half of the aircraft was missing, having burnt to ashes on the forest floor. Major Armour's team had made some progress in their analysis. The cockpit T-handle for the No. 2 engine was retrieved from the wreckage in a position that immediately spelled trouble. To battle an engine fire, the handle is supposed to be pulled out — cutting fuel and oil flow to the engine — then twisted left and right about seventy degrees, triggering two fire extinguishers. But the T-handle was found in the wreckage pushed part-way back in, which would have restored the flow of fuel and oil to the dangerously hot engine compartment. Indeed, the laboratory in Hull, Quebec, reported that the oil valve in the No. 2 engine was found in the open position (the companion fuel valve was too burned to determine its final position). It seemed clear the T-handle had become untwisted and inadvertently pushed back in again, electrically resetting the fuel and oil systems. But how had this happened?

Armour's team first determined that a U.S. Navy ground simulator for the Labrador, used to train Canadian pilots, had a T-handle that did not twist. Rather, the engine fire extinguishers on the American version of the Labradors were triggered by a separate toggle switch. Thus Canadian pilots may have subconsciously learned a bad lesson: that the T-handle's "normal" position is untwisted, even when battling an engine fire. Next, Armour visited Labrador squadrons in Greenwood and Comox, asking the pilots to demonstrate to him their engine fire-fighting procedures. Sure enough, about half of the pilots centred the T-handle after twisting it left and right, putting the handle in a position that could allow it to be accidentally pushed back in and thus restore the dangerous oil and fuel flow. New instructions were quickly spread throughout the rescue squadrons: leave the T-handle in a twisted position to prevent the inadvertent resetting of the valves. There was no evidence that the pilots of Labrador 305 had actually pushed the T-handle back in; the job might just as well have been accomplished by the excessive turbulence or by aircraft vibrations.

Inspections of Labrador engines across the fleet also showed a potentially dangerous kink in a fuel line. The double-shielded line, unlike many parts that are assigned a lifespan and replaced periodically, was regularly inspected and replaced only if technicians spotted chafing, wear or splitting. There was no proof from the damaged No. 2 engine that the suspect fuel line had fed the engine fire for Labrador 305, but the air force nevertheless decided to replace it with triple-shielded lines on the remaining twelve aircraft. Armour suspected Labrador 305's fuel line was slowly seeping gas fumes that had fed the initial fire and caused the fire to re-ignite suddenly when the T-handle was pushed back in. Could that sudden re-ignition have been the source of the explosion that apparently blew the aircraft apart?

The only way to test the hypothesis was to recreate the blast. And so the munitions testing facility at Canadian Forces Base Suffield, Alberta, was asked to determine how big an explosion could result from the sudden ignition of a small quantity of gas inside an engine compartment. Technicians atomized about 500 millilitres of gas with air inside a steel tube about two metres in diameter. At one end, a piece of plywood sealed the chamber. At the other end, technicians placed a simulated engine panel. They then triggered a blast, filming the results at high speed. The explosion — which consumed only about 250 millilitres of fuel, leaving the remaining 250 millilitres as residue — was measured at between eight and nine pounds per square inch, a respectable force but nowhere powerful enough to split a Labrador in two. Nor did it damage some Canadian Tire tool boxes that the technicians had placed inside the tube to see whether the badly dented metal boxes from inside Labrador 305 had been damaged by an engine compartment explosion. The blast, however, did knock out the mock engine panel at one end of the steel tube. The Suffield results thus matched the damage on the right engine compartment cover found in the Marsoui wreckage. The plate, which had torn free of its fasteners on the ceiling of the passenger compartment, clearly was blasted onto the floor of the aircraft but had not caused any structural damage. Fleet-wide maintenance records helped to confirm the hypothesis; they frequently noted a fuel residue on the engine panels when they were opened after engine operations.

The most unexpected finding from Suffield lay in the film of the blast. Played at normal speed, it looked like an unexceptional explosion. But when the film was slowed down, investigators could see a long finger of flame — perhaps eight metres long — shoot out from the tube in a flash. Up to this point, Armour had reconstructed a likely sequence — a fuel-line-fed engine fire that was extinguished and that re-ignited explosively — yet he had discovered nothing powerful enough to cause the breakup of the Labrador. However, the bright ribbon of fire that probably darted out of the engine compartment might be the link to the big external explosion that most of the witnesses reported. But where to find the fuel to feed the fireball at the rear of the helicopter? And even if there was such a fireball, how could it break up the fuselage? Indeed, it was clear the rotors had cut into the aircraft. Had there been two separate events?

The team was by now exhausted and hitting too many brick walls. The Christmas holidays were coming, almost three months had passed since the accident, and they still had no explanation for the disaster. Jim Armour's group had worked flat out until October 27, when the flying restrictions were finally lifted, and their wives insisted they slow down. But the pressure remained, and they spent many long days trying to come up with solutions that would explain the strange, blackened array of wreckage assembled on the jig. Their small office in Hangar No. 5 was dubbed the "dead end" room as every theory seemed to founder. An easel in the room held large sheets of paper that contained numerous abandoned scenarios. "We were literally exhausted pre-Christmas," Armour recalled. "The boys pulled me aside and said, 'You need to walk away for about a week.'" And so he did, recuperating and resting from an intense three months of work.

Christmas came and went, and on January 8, Armour and his chief technical investigator, Major François Tremblay, were back in the "dead end" room trying to come up with solutions. "Frank and I sat in that room thinking, there's something that's not right," Armour recalled. Labrador 305 had a professional, experienced crew. They had been through in-flight emergencies before. They would have been pro-actively dealing with the Marsoui engine fire, dousing the blaze, picking out a landing spot, and — hell — dumping fuel. A Labrador on only one engine needs to lighten up quickly, and the only fast way to do that was to dump fuel. But the wreckage indicated that the fuel-dump tubes had not been extended. "We were sitting around the room saying, 'I just can't understand how this would be,'" Armour recalled. Then came the flash of inspiration. Photos. Check the photos. Could the fuel-dump tubes have been reset inadvertently by one of the investigators when the wreckage was being recovered? Perhaps they had been pushed back in by someone unaware of the importance of this clue. The photos might at least show how the tubes appeared when they hit the ground.

Tremblay and Armour gathered the photos together and arranged them according to a chronology, from the first pictures of the wreckage to those taken many days later. The earliest pictures had, in fact, been snapped by a news photographer who had been given privileged access to the crash site by the Quebec police force because there was no police photographer immediately available, and the police would need these pictures for their own official record of the disaster. A Quebec newspaper was later justly criticized for publishing some of these photos, which included images of the dead bodies of the Labrador 305 flight crew. But two of the pictures inadvertently recorded a key detail that finally broke the investigation logjam. There, in one small section, clearly visible, was at least one of the fuel-dump tubes fully extended. Indeed, a photo taken by a military photographer not long after showed the same fuel-dump tube extended. "Right away, we got really excited," Armour said later. "It was kind of a *eureka*. I can remember at the end of the day phoning up [his boss] Colonel Legault and saying, 'I think we've got it.'" The pictures demonstrated that the crew had indeed been dumping fuel as expected. And the discovery finally provided a potential fuel source for that huge, external explosion that all the witnesses reported. The earlier results from Suffield, showing the long lick of flame flashing out from the engine compartment, suggested the means by which that fuel-dump had been ignited.

The next day, Armour called technicians from the military lab in Hull, Quebec, who came to the hangar to retrieve the two fuel-dump tubes to try to determine from physical evidence whether they had indeed extended. The answer some days later was that yes, one had extended its full length, confirming that the crew was dumping fuel as expected. But the other one, on the right, had failed to extend, allowing the fuel to emerge uncomfortably close to the right-hand side of the fuselage. Armour's team then gave themselves an intense course in fuel-dump technology, reviewing manufacturer guidelines back to the early 1960s and following the Canadian military's evolving policies on fuel-dumping throughout the next three decades. As a result of Armour's discovery, on March 8, 1999, the air force ordered inspections of the fuel-dump tubes throughout the remaining fleet. Technicians eventually found that nine of twenty-two tubes inspected did not extend as intended, demonstrating that there was indeed a problem. Informal checks among the squadrons showed that the Labrador crews were well aware that the tubes did not protrude reliably when needed, though the failure was never considered a critical problem.

Soon after the January 8 epiphany in Hanger No. 5, Armour requested that the military conduct a filmed recreation of a Labrador fuel dump so that the Marsoui witnesses could help determine whether that was what they had seen. Later that month, a crew from 424 Squadron at Trenton carried out a Labrador fuel dump over a little-used military strip at Mountain View, Ontario, from an altitude of about 500 metres. Three ground-based video cameras recorded the sequences — three dumps in ten-second bursts. The experiment showed that even with the fuel-dump tube unextended, the fuel cleared the fuselage fairly well. During the week of February 8-11, the videotape as well as a desktop model of the aircraft was shown to the six most reliable witnesses to the Marsoui disaster, including one man who had not been interviewed by anyone previously. This person had tied his dog to the front door of his house on October 2 to keep the media away, thus escaping the attention of the military as well as reporters during the first days after the crash. He was found in February and, with the others, indicated that the fuel-dump at the time of the crash seemed to match the videotape.

The Suffield experts were again asked whether dumped fuel could create a major external blast with enough power to at least alter the trajectory of the rotor blades. And Boeing Vertol, based near Philadelphia, Pennsylvania, was asked whether an external blast could have such an effect. The Suffield experts conducted tests that only suggested the effect Armour's team hypothesized — they could not provide definitive numbers. Boeing likewise could not say definitively that an external blast would alter the course of the rotor blades, in the absence of experimental results from Suffield. They could only say that it was possible. The Boeing experts did advise, however, that the Marsoui crash was the first time such an explosive accident had happened to the any of the twin-rotor helicopters the company had ever manufactured.

Armour and his team concluded that the accident was the result of an extraordinary chain of incidents, the origins of which were not fully understood. The engine fire was likely the result of a weakened, leaky fuel line. The failure to douse the fire probably occurred because part of the blaze had migrated to a section of the engine compartment not reachable by the extinguishers. The likely re-ignition was the result of the unintended resetting of the T-handle. The resultant engine-compartment blast probably created a long lick of flame out the back that ignited the dumped fuel, knocking the rear rotor blades out of alignment. The aircraft was then chopped apart and fell to the ground in pieces, one of its fuel tanks exploding in the process. No individual element of the hypothesized sequence could be proven definitively. But the scenario was the best match Armour's group could make with the known facts of the disaster.

The only certain way to avoid similar accidents was to break several links in the chain. The fuel lines for the engines, for example, had all been replaced with triple-shielded versions. Zero-tolerance inspections from now on would not pass any lines that were even slightly frayed. Pilots were warned to keep the T-handle twisted when they fought engine fires to prevent any inadvertent resetting of fuel and oil flow.

Engineers started to redesign the extinguishing systems to disperse firefighting chemicals throughout the engine compartment so flames could not hide in any far corner. The fuel-dump tubes would be inspected every fifty airframe hours rather than every 440 hours to ensure they would extend properly in an emergency, and the air force reviewed its fuel-dumping techniques to ensure that pilots did not carry out the procedure in conditions that kept the fuel too close to the fuselage. "Should the same most probable sequence of events line up," Armour wrote on May 4, 1999, "the chain would be broken before a recurrence." The crash of Labrador 305 was also instrumental in a January 2001 policy change that requires every new aircraft purchased by the Canadian military to be equipped with a cockpit voice recorder and flight data recorder. Future investigators would not have to face the same hurdles as Armour's team, and nervous aircrews would have their answers much sooner.

A year after the tragedy, the Labrador 305 "tin-kickers" — Jim Armour, François Tremblay and Brian MacDonald — gathered in a cavernous hall deep in the bowels of a giant military testing facility in Hull, Quebec. The room is the size and height of a gymnasium, a windowless warehouse with rough metal shelving along two walls. Scattered like poorly presented museum pieces are the remnants of Canadian military aircraft mishaps and disasters. On the floor is the battered tail of a T-Bird aircraft, a CT-133 Silver Star. Next to it is an enormous CF-18 engine retrieved from the wreckage of a takeoff flop in the Arctic. Suspect ejector seats clutter the middle of the concrete floor. Along the walls are cardboard boxes, roughly labelled and filled with the twisted metal remains of other crashes. A skid on the floor and a few big cardboard boxes hold the bits and pieces of Labrador 305 that were retained after the reconstructed airframe was finally taken down from its jig in Hangar No. 5 at Uplands. Here are both the engines, neither of which shows any sign of trouble. The rotor heads are here as well, along with the bright-yellow lower half of the search-and-rescue door from the right side of the helicopter just behind the bulkhead separating the cockpit from the passenger cabin. A sign on the skid containing the rotor heads reads: Quarantined. The front rotor head would be sent eventually to 413 Squadron in Greenwood to serve as part of a military memorial.

Surrounded by this disturbing wreckage, Armour and his team can still speak admiringly of Labrador helicopters. They have come to regard Labrador 305's tragic end as almost freakish, a bizarre series of mishaps that would be difficult to repeat even without the aircraft improvements that emerged from the investigation. A million-to-one chance, they say. "I was extremely worried about the aircraft," Armour said of his feelings at the start of the Marsoui investigation. "We'd had a couple of fairly ugly incidents with it . . . it's getting older and it has had leaks. It did cross my mind that, hey, maybe there is a big problem with this airplane. But the more we looked at the way the thing is designed, put together, the safety measures that are there, in reality it's very well engineered. By the end of the investigation, I had a lot more respect for the aircraft. I had a very good opinion about the aircraft." The Labrador's systems can take a lot of abuse and still fly safely, Armour says, despite their advanced age. Labrador 305 was simply overwhelmed with too many unforeseeable problems at once.

The search-and-rescue technician who was the first to come upon his dead comrades at the crash scene in Marsoui quit the military less than a year after the accident. Maurice Robert moved with his wife and two young children to New Brunswick, ending a twenty-one-year career with the Canadian Forces in July of 1999. Known as Moe to his 413 Squadron buddies, Robert had been thinking of getting out even before Labrador 305 went down. "I might have stayed a little longer but I guess the straw that broke the camel's back was the fact of that crash," he says. "I miss it a lot, as far as the work and the comradeship are concerned." What Robert doesn't miss, he says, is the callousness of the military, never more evident than immediately after the crash. "While they didn't have any information, right away they're saying the aircraft is safe. That's the thing initially I had a problem with. Initially they said, okay, you can go flying right away." Only after crews got together to collectively convey their worries did the military hierarchy agree to delay the squadron's return to flying.

Robert's wife had been a close friend of the wives of the men killed, and the tragedy weighed that much more heavily upon her. Her concerns about her husband's safety on the Labradors helped tip the balance in the family's decision to get out of the rescue business. Ironically, on the very day the squadron was to fete Robert and his wife with a farewell dinner, there was another Labrador incident. On May 20, 1999, Robert was training in a Labrador on the south shore of Nova Scotia's Minas Basin, at a spot ominously known as the Black Hole, when a blade of the helicopter clipped a tree, forcing an immediate precautionary landing. The Labrador crew was forced to wait several hours for ground technicians to arrive by truck, and Robert was late for his retirement dinner — causing a last few anxious moments for his long-suffering spouse.

■ ■ ■

Jeremy Tracy, a tall, lean Brit, is a helicopter test pilot who has been grounded by good weather on this mild winter day in January 2000. His prototype helicopter, designated PP-7, is tucked inside a cluttered hangar at the Shearwater air base near Halifax. Tracy and his crew are in Nova Scotia to test the anti-icing capabilities of the Cormorant, the high-tech aircraft that, beginning in late 2001, will replace the Labradors. Tracy has spent his aviation career trying to avoid poor weather, but the Cormorant needs to be flown through icing conditions to demonstrate its ice-fighting technology for certification in Canada and elsewhere. He and his four crew scramble into the air whenever there is a fierce snowstorm in the Halifax area, chosen because it presents some of the worst icing conditions that the world has to offer. Labradors cannot fly in icing conditions, but the Cormorants can — thanks to heated rotor blades, heated air intakes for the engines, and a smooth fuselage designed to repel ice. On January 21, a severe snowstorm virtually closed the Shearwater airport, but Tracy took PP-7 aloft twice, for an hour each time, to check the anti-icing capabilities. "We're the only guys in Halifax who have a big, broad grin on our faces when we see foul weather," he says. "It's just the thing we're looking for." Video cameras mounted on the rotors and elsewhere on the fuselage record how ice is formed and shed. All the crew wear parachutes in case of trouble, but the tests show the blades, engine intakes and fuselage shedding ice as intended.

Today, however, the weather feels like spring. The Cormorant team must spend the day inside a spartan office trailer reviewing videotape while the crews of several Sea King helicopters nearby jump at the chance to fly in good conditions. Unlike the Labrador, the Cormorant has autopilot and an anti-vibration system for a smoother, passenger-friendly ride. Crews can arrive at rescue scenes well rested and sharp after a long transit. The de-icing system allows safe flights in the kind of weather in which rescues typically take place. The Cormorant's standard range is 1,018 kilometres compared with about 835 kilometres for the Labradors, and they're designed with the potential to accept auxiliary fuel tanks under the floorboards that would extend that range much farther. They cruise up to thirty per cent faster than the Labradors, and the auto-hover system makes rescue hoisting safer.

The Cormorant also uses three 2,000-horsepower engines as a fail-safe measure, allowing it to climb "like a homesick angel," as one Canadian rescue pilot put it. The single-rotor aircraft can fly quite easily when one engine quits — the remaining two engines automatically power up to take over the load. Indeed, the manufacturer advises that crews can actually save fuel by deliberately shutting off the third engine during a long cruise. Even in the unlikely event that two engines quit, the aircraft can recover on just one engine, thanks in part to extra-wide rotor blades that provide strong lift. The Cormorant's fire-fighting system is far more sophisticated than the Labrador's. There is no T-handle but rather a series of buttons and sensor readouts on the instrument panel. A Cormorant losing an engine to a fire would not need to go to ground immediately, would not even have to slow down, and would not need to dump fuel. If the crew of Labrador 305 had been flying a Cormorant, losing an engine would have been a minor blip on a relaxing flight back to base, with the aircraft virtually flying itself through the high winds. Such a helicopter would have passed briskly over the picturesque Chic-Chocs, its crew resting comfortably, their thoughts drifting toward family and friends waiting expectantly at Greenwood.

In 1984, the same year the *Ocean Ranger* Royal Commission criticized the Labrador helicopters as "unsuited for marine rescue operations offshore," the Defence Department launched a project to replace the fleet with modern choppers. The Labradors were acknowledged to be slow, unreliable, unable to fly in icing conditions, lacking in night-vision technology, and inadequately equipped with communications and navigation gear. The procurement process would be complete by 1995, according to the project team, when the last of the Labradors would be retired. Three years later, with no contract in sight, the project was changed: the military would buy new search-and-rescue aircraft — helicopters and planes — in a single order. The changed 1987 framework, which took the focus away from the aging Labradors, sounded ominous to the helicopter community because it tied the fate of their obsolete machines to other priorities. Later that same year, this new plan was abandoned as the Defence Department decided to expand its entire Hercules fleet and to continue using some of the four-engine transports for search-and-rescue work. Due to the resulting delay in the helicopter replacement program, in 1989 the military had to extend the life expectancy of the Labradors to 1998 and institute a careful maintenance program. The "airframe has been in service since 1963, and will be thirty-five years old by the time it reaches its current ELE [estimated life expectancy] of 1998," an internal military analysis noted. "Life extension beyond this would involve substantial risk." In 1990, the federal government decided to buy a single-model helicopter to replace both the

A Cormorant helicopter during icing trials at CFB Shearwater, Nova Scotia. DB

Labradors and the sub-hunting Sea King helicopters, once again tying the fate of the rescue choppers to other defence priorities. On July 24, 1992, the Conservative government of Prime Minister Brian Mulroney finally announced a plan to buy fifty EH-101 helicopters, fifteen of which were for search and rescue. Upgrades on both helicopter fleets had been put in abeyance since the 1980s in anticipation of the contract and the imminent arrival of new, modern aircraft; nevertheless, the life expectancy of the Labradors was now extended for a third time, to 2001, to accommodate delivery of the last EH-101.

The announcement acted as a lightning rod for political dissent. Peace groups decried the $5.8-billion price tag as an extravagant waste during a time of fiscal restraint or as old-fashioned warmongering despite the end of the Cold War. Liberal Leader Jean Chrétien set about demonizing the "Cadillac" EH-101. "We're not in the Cold War any more, and we need that money for Canadian problems," he said. Lost in the debate was the strictly humanitarian purpose of fifteen of these helicopters, which would be used solely to rescue civilians inside Canadian territory. The EH-101s had become a political football, and, recognizing the electoral danger, the new Conservative Leader Kim

Campbell cut the order to forty-three aircraft, saving a billion dollars by eliminating seven of the sub-hunting models. The move, coming just before the fall 1993 federal election campaign, did nothing to improve Conservative fortunes, and Liberal Prime Minister Jean Chrétien cancelled the deal within hours of being sworn into office. The contract penalties and other costs — at least $500 million, and likely much more — would have covered most of the invoice for new search-and-rescue helicopters.

The Labrador search-and-rescue community was now operating in the worst of all possible worlds. Every effort since 1984 to replace the helicopter had been scuttled by foot-dragging and politics, there was no new helicopter on the horizon, and important upgrades — including reliable engines — had all been cancelled. The airframe's life expectancy was pushed ahead once more, to 2003, partly by planned "cannibalization" — stripping precious spare parts from the first aircraft to retire to keep later ones operating. The long-overdue engine-upgrade program was finally approved in 1996 and was completed three years later. Even with new engines, however, the Labradors often sat unserviceable or unable to reach distant or weathered-in rescue sites.

Labrador 305 almost certainly would still have been in service even had the Liberal government allowed the EH-101 order to proceed as scheduled in 1993. The first deliveries were scheduled for 1998, though production glitches typically unravel the best planning. But Labrador 305 should have been retired far earlier, by 1995 at the latest, were it not for the maddening inertia of government. The military's search-and-rescue community had for years silently coped with the crumbling capital inheritance of their Labradors and worked with tenacity, ingenuity and professionalism to overcome many of the technological failings of the aircraft. With no union and under gag orders, they lacked a voice to make their case a matter for informed public debate. The crassly partisan decision of the Liberals to cancel the replacement project and then to procrastinate for years before ordering a new fleet allowed the sweat and toil of its search-and-rescue specialists to mask its own irresponsibility. Labrador 305 may have been a "robust" aircraft design, as the crash investigators claimed, but it harboured mechanical gremlins that even the best maintenance crews in the world could readily miss. Sticky fuel-dump tubes, leaky fuel hoses, unreliable fire extinguishers, cockpit controls with no fool-proofing — these all ganged up on an outstanding crew in the fall of 1998, when Labrador 305 should have been on display in an aviation museum.

Canadians were reminded of the aging Labradors during the search in March 2001 for three teenage boys who died at Pouch Cove, Newfoundland, while hopping from ice floe to ice floe. Front-page newspaper photos and national TV news footage showed the distinctive bright yellow-and-red fuselage hovering close to the shoreline as the crew searched the ice-choked waters for bodies. The sight of these pounding, ungainly-looking machines can often inspire a confidence among ordinary Canadians that, no matter how terrible the disaster, they are well-served by Ottawa's search-and-rescue regime. The valiant crews inside know better.

Chris Flemming aboard CCGS *Sambro*. DB

Rescuing the Future

Chris Flemming can name every rock, cove, bay and point along this morning run out of Sambro, Nova Scotia, as well as every Cape Islander fishing boat he encounters. He can readily identify the owner of every multicoloured buoy marking one of the hundreds of lobster traps draped along the sea floor below. Houses perched all along the rocky, sea-worn cliffs in the distance belong to Bill, or Stuart, or Chuck, who built himself an addition last year. Flemming can also describe the geography of the ocean floor, naming every hidden shoal and channel, describing the bottom as rocky or sandy or gravelly. And every feature he points out seems to have its own story, usually uproarious, which Flemming tells with good cheer in his broad south-shore accent. He can also read the water itself, the tidal flows and currents, the sea state. And he can read the air, the breezes and gusts and cloud formations, especially watching for signs of the easterly gales which are, he says, "rank poison." Flemming's close knowledge of these inlets and islands and of the grey, brooding Atlantic beyond draws on his twenty-three years in the coast guard and his many years before that as a lobster fisherman. His mind is a living chart cabinet. He has a deep, visceral understanding of the moods and habits of the sea.

Flemming is piloting the CCGS *Sambro*, a 16.25-metre cutter named for its home port, a fishing village about a half-hour's drive southwest of Halifax. The red-hulled *Sambro* shares the harbour with about forty fishing boats that head out almost year-round for tuna, shark, halibut or lobster. The *Sambro* was designed and is operated as a state-of-the-art rescue craft, one of only a few vessels in the coast guard largely exempt from the fleet requirement for multi-tasking, that is, performing fisheries patrols, environmental watches, buoy tending and a grab-bag of other duties. The word "Rescue" and its French equivalent "Sauvetage" are painted in large black letters on the scrubbed outer white walls of the wheelhouse. Designed in Britain and built by a Lake Erie firm based in Wheatley, Ontario, this spunky little boat — known as an Arun-class vessel — is just three years old. Two diesel engines give the Aruns a top speed of twenty-four knots, but the aluminum-hulled *Sambro* is so loaded down with rescue gear that it can reach only about twenty knots full out, or twenty-one knots with a following sea, as Flemming observes. Early models were built with fibreglass hulls but were prone to waterlogging; sea water could seep inside and saturate the fibreglass matrix, adding as much as two tonnes to the weight and thereby reducing speed. The coast guard Arun based in Port Bickerton, Nova Scotia, is one such hull-soaked model, but the younger *Sambro* has benefited from the problems of its predecessors.

The $2.5-million cutter is a highly manoeuvrable vessel. It can turn around in its own length when Flemming puts one of its two propellers in reverse, and it is highly responsive to the wheel. Flemming recalls once cresting a giant wave perhaps three times higher than the boat, and as the *Sambro* shot down the other side it became briefly airborne, then dove nose first under the trough. It popped up again with no more damage than a lost windshield wiper and a minor piece of the mast gone. No boat is unsinkable, but the Aruns have a better chance than most. The half-dozen watertight compartments filled with plastic bubble foam mean that, even if the hull is gored, very little water can enter. Indeed, the vessels are designed to flip upside down and quickly right themselves, something that would never be done deliberately but that has happened more than once during training sessions gone awry. Flemming sometimes complains about the short spacing of the stairs leading below deck, as well as the low ceiling of the pilot house, blaming it on short-statured English engineers with short legs. At least the ceiling is fitted with grey padding for the inevitable bumps and jolts.

Flemming and his three crew members work seven days on, seven days off, and they remain on call by pager twenty-four hours a day when they're on watch. They must live close enough to the *Sambro* that they can get the boat under way in thirty minutes or less whenever the call comes from the Halifax rescue co-ordination centre. Most of the time, the crew can head out to sea in fifteen minutes, and in just five minutes between eight a.m. and four p.m., when they're required to be stationed dockside. For the seven-day stretch when they're on call, they have to forgo movies, alcohol and any trips beyond a thirty-minute radius of

CCGS *Sambro*. DB

Sambro. "We have to be good guys for seven days," says Flemming. The vessel has a small Zodiac boat strapped to its deck, and it shares the wharf with a fast rescue craft, a rigid-hull inflatable boat that can hit thirty knots at top speed and can work in tandem with *Sambro* if necessary. *Sambro* was the first search-and-rescue boat on scene during the 1998 Swissair disaster, making Peggy's Cove that night in about fifty minutes flat. Its territory, though, extends much farther, down the south shore of Nova Scotia for about 100 kilometres and another 100 kilometres to the northeast, then out to Sable Island near the extreme limit of its 125-nautical-mile range, when it has to turn around for home or face running out of fuel. It's a large, forbidding territory but one that Flemming knows well.

■ ■ ■

The Canadian Coast Guard has been badly demoralized ever since the April 1995 transfer of its fleet from Transport Canada to Fisheries and Oceans — a "hostile takeover," as many embittered coast guard employees put it. The move was made to save money — as much as $25 million a year — as the Liberal government of Prime Minister Jean Chrétien wrestled with persistent federal deficits. For many of those inside the service, the merger remains as traumatic as the forced unification of the three branches of the armed forces in 1968. Service pride eroded as the coast guard found itself subsumed into the broader fisheries priorities, and its traditional expertise in search and rescue was devalued. Already considered the junior service in the military-run rescue co-ordination centres, the coast guard members now seemed a mere adjunct to the Fisheries Department. But the assault on the coast guard in fact had begun much earlier. The decline really began in 1991, with an extraordinary analysis that attempted to determine whether the number of lives saved by the coast guard was really worth the high cost.

The October 1991 report, commissioned from the consultants Sypher: Mueller International Inc., was the first top-to-bottom analysis of the coast guard's rescue capabilities in nine years. The study examined detailed statistics for three years of marine incidents in Canada. The authors found that the coast guard was directly responsible for saving the lives of about 110 people each year, or about seven per cent of all the lives that had been at risk. Drawing on a series of Canadian government-sponsored research studies, the report determined that the generally accepted value placed on the average human life ranged between $200,000 and $2.5 million, based on such things as lifetime earnings potential, court judgments and insurance settlements. However, the authors argued, many of the people rescued had voluntarily placed their lives at risk by undertaking inherently dangerous activities such as fishing. The "voluntary" aspect of their predicament tended to lower the assigned value of individual survivors' lives to perhaps $750,000, said the report, without further elaboration. For the period studied, the coast guard spent about $492,000 for each life saved in marine incidents, or about $430,000 for each life saved in all rescue incidents, including air crashes and medical evacuations. The cost-benefit ratio thus worked out to 1:1.5 and 1:1.75, that is, the value of the life saved was 1.5 times or 1.75 times the coast guard money spent to save it. And when the value of the property saved was factored in as well, the cost-benefit ratio improved to almost two-to-one. Good value for the money, in other words.

The authors went on to analyze the types of coast guard vessels used to save these lives and found that all but the largest ships proved cost effective, that is, they saved lives valued well in excess of the cost of saving them. The glaring exceptions were the 600-Class ships, the biggest in the coast guard fleet, none of which had saved a single soul in the period from 1983 to 1987. The analysis determined that their cost-benefit ratio was perhaps 1:.32, that is, their rescues typically expended resources worth about three times the life of the hapless survivor plucked from the salty swell. The report predictably called for an end to 600-Class rescue operations, and it suggested ways to further improve the cost-benefit ratios of the remaining fleet. Better use of boats from the coast guard auxiliary, for example, could save tax dollars as it has in the United Kingdom, where the volunteer Royal National Lifeboat Institution operates 200 lifeboat stations funded independently of government. Safety promotion and better enforcement of regulations, as well as cost-recovery measures, could also keep the lid on costs, the report said.

A little more than a year later, the federal auditor general also weighed in on whether the coast guard's search-and-rescue program was cost-effective. Drawing on rescue statistics for 1989 and 1990, the auditor general's office found that dedicated rescue boats played a key role in only about twelve per cent of all marine distress incidents. "Our review has indicated that, although the federal search and rescue resources were often called upon first in search and rescue incidents, they did not carry out the actual rescue for most distress incidents, because other resources closer to the scene were able to." The study also found that the search-and-rescue fleet spent only about three per cent of its available time responding to incidents, and some boats had not had a single rescue in years. The auditor general's most startling recommendation was that the coast guard should consider simply decommissioning its entire rescue fleet and end its rescue patrols, relying instead on volunteers and other government ships as needed.

The coast guard was already in the process of restructuring its fleet when these outside voices gave credence to the federal government's insistence on cost-cutting and downsizing. After the election of the Liberal government in 1993 and another study on ship management the following year, the coast guard found itself ordered to merge its fleet with the Fisheries and Oceans fleet. "Although the total number of vessels employed may be reduced somewhat as a result of efficiencies of scale, multi-tasking of vessels could enhance the efficiency and possibly the effectiveness of marine search and rescue," Bill Slaughter, the head of the military-run National Search and Rescue Secretariat, said in January 1995 in support of the move. Money soon began to disappear rapidly. Fisheries and Oceans spent about $163 million on search and rescue in the first full year of the merger, then cut millions out of the budget each subsequent year to an estimated $109 million in 1998-99. On paper, at least, the Chrétien government appeared to be reaping program efficiencies of more than $40 million a year.

The reality was quite different. Search-and-rescue services suffered badly, and overworked staff often had to make up for the shortfalls. In the case of the crash of Swissair Flight 111, a terrible disaster finally exposed a festering manpower problem at the Halifax rescue co-ordination centre and forced officials to double the number of marine controllers. The 1998 *Flare* disaster also demonstrated how service had deteriorated. In its final report on the sinking, the Transportation Safety Board of Canada criticized the coast guard for failing to have a large ship stationed along the south coast of Newfoundland. Instead, downsizing had left the area with only two Arun cutters, neither of which was suitable for offshore incidents involving large commercial ships. Before the cuts, the coast guard had always had a buoy tender working the south coast that could be readily diverted to search and rescue as needed. Marine controllers had no such option in the *Flare* case; they had to call in the CCGS *Ann Harvey* from Newfoundland's west coast, more than fifteen hours away. "With budget cutbacks, secondary resources are now stretched a lot thinner, and those secondary vessels are not always in the area," says a board report. "There has been a fairly drastic reduction in SAR [search and rescue] coverage in most of the areas within the Newfoundland Region." The report warned that the thin coverage along the south coast, a busy corridor for cargo traffic bound for the St. Lawrence Seaway, appeared to violate international search-and-rescue agreements signed by Canada.

Another safety board investigation demonstrated how cost-cutting had placed lives at risk. A February 1999 report into the near-sinking of the scallop dragger *Scotia Gold* determined that a coast guard helicopter had lowered a faulty, unworkable pump to the struggling crew. This gas-powered Honda pump was made of steel and an aluminum alloy, a combination readily corroded by salt water. Indeed, the user manual carried the explicit warning: "to avoid pump corrosion, never pump sea water." The board determined that the Honda pumps were much cheaper than the Briggs & Stratton salt-impervious pumps that the coast guard had used previously, but the coast guard had replaced them with about thirty of the Honda models "primarily as a cost-cutting measure." The embarrassing report forced an end to the cheap-pump program and a reversion to the more reliable — albeit more expensive — model.

The most humiliating lapse occurred in St. John's when a coast guard ship was pulled from search-and-rescue duties to conduct an evening cruise for twenty-two bureaucrats from Corrections Canada. The light icebreaker *J.E. Bernier* was used to host a two-hour dinner party on September 15, 1999, featuring lobster, bacon-wrapped scallops and tiger prawns. The entire bill — $9,820.37 — was picked up by the cash-strapped coast guard, while the offshore patrol vessel *Cape Roger* was ordered to stand in for the *Bernier*'s scheduled search-and-rescue watch that night. A suspicious coast guard auditor investigated the next day and found several violations of regulations, including an apparently deliberate failure to record the event in the ship's books. To many of the men and women in the coast guard, this was galling proof that the years of cuts and the forced merger had diluted their high principles and traditions of service. The federal fisheries minister, Herb Dhaliwal, promised that disciplinary measures would be taken, though he refused to provide details, and that new policies would prevent a recurrence.

The Canadian Coast Guard's troubles thus parallel those of the military's search-and-rescue squadrons, though without the high public profile of a Labrador crash. The coast guard's rescue resources were badly diminished throughout the 1990s, leaving the service often unable to respond with dispatch. The Swissair and *Flare* operations had both briefly exposed the coast guard's ineffectiveness and led to a few modest improvements. But as search-and-rescue veterans in the service readily acknowledge, the coast guard's long-standing problems will not be resolved by a patch here and a bandage there. "Coast Guard, departmental senior management and the Minister must recognize that Coast Guard is poised on the brink of an abyss," a top official had warned in the wake of the Swissair disaster. Little has changed since that alarm was raised.

■ ■ ■

Chris Flemming brings the cutter *Sambro* as far as Portuguese Cove, along the west entrance to Halifax harbour, before heading back home for lunch. All of the boat's systems are in working order. A giant cargo ship appears far off to the west, and Flemming mentions in passing that we'll be hearing from it shortly. Seconds later, the captain's voice can be heard crackling through on Channel 14, identifying himself to the coast guard radio station in Dartmouth and declaring his intention to enter the harbour. Minutes later, we come upon a lobster boat winching up traps and escorted by a blizzard of hungry gulls. Flemming predicts the catch will be poor, and sure enough the traps contain the odd sculpin and flounder but few lobster. This late in the season, the colder water drives the crustaceans farther offshore. Flemming is like a cop on a familiar neighbourhood beat; not much gets past him. His long experience of these waters and his knowledge of the vessels that ply them provide search and rescue with its vital link between past and present.

The *Sambro*, on the other hand, is a symbol of the future — one of the few bright spots in the coast guard's struggling search-and-rescue service. Flemming at his wheel is surrounded by glittering technology. On the left is a large colour monitor and a computer that can access an electronic library of 166 charts for these waters. Keyboards, mouse pads and a CD-ROM drive are all within easy reach of the captain's chair to help him manipulate the data in the system, technology that was added after *Sambro*'s 1997 delivery. An orange axe-head shape superimposed on the charts marks the precise position of the *Sambro*, drawn from the satellite-based Global Positioning System that constantly updates the vessel's longitude and latitude. To the right is another large monitor, a colour radar replacing the $30,000 radar that came with the boat and was soon chucked. At the rear is another monitor that provides a complex sonar-based picture of the sea bottom and everything in between. Behind that is a laptop computer that sends a regular signal on VHF radio or satellite to the nearest coast guard radio station about the precise location of the *Sambro*, also drawing on GPS data. The boat's own radio systems are supplemented by an ordinary cellphone that Flemming always keeps close at hand. Down below deck is a sophisticated self-locating datum marker buoy that can be dropped at an accident site to track the drift of survivors by satellite.

The *Sambro*'s spirited crew is also a harbinger of the future. These Arun-class cutters were designed in Britain to carry a crew of seven, but for the Canadian version the crew complement was reduced to five, and, accordingly, five chairs were installed in the wheelhouse. However, lean times at the coast guard have further reduced the Arun crews to four — a commanding officer, an engineer and two deck hands, at least one of whom has medical training as a rescue specialist. Thus while more technology has been added, the empty chair stands as a reminder of the broader trend in search and rescue to operate with fewer professionals. These smaller crews, at the same time, must be more intensely trained to handle the evolving electronics. The *Sambro* crew points the way ahead in another important respect: one of the deckhands is a woman, Norma Scarfe, a nine-year term employee and the only female lifeboat crew member in the Maritimes. Search and rescue is still an overwhelmingly male profession but this will inevitably change as ability becomes the fundamental criterion.

The *Sambro*'s logbook witnesses the larger changes taking place in search and rescue in Canada. Disabled or foundering fishing vessels once dominated the station's rescue work. They still represent much of the caseload, and fishermen frequently embed hooks deep inside their flesh. But in recent years, more maydays are coming in from pleasure craft, especially the polished yachts that berth in the clubs at Chester and Mahone Bay and along Halifax's Northwest Arm and Bedford Basin. This sector is growing rapidly as wealthy baby boomers retire. Safety standards are often lax, as eager novices head out to sea with little more than a cellphone for communications. *Sambro* answers about 110 calls for help each year, up from its former annual caseload of eighty-five. Most of that new work comes from pleasure craft, which now account for about half of all rescues, most of them in the summer and early fall at the height of the tourist season. Flemming has also recently taken *Sambro* out to search for a woman tourist washed off the rocks at Peggy's Cove, a recreational diver hit with the bends and a sea-kayaker who failed to check in at the promised time. The booming outdoor-adventure sector will place increasingly heavy demands on search and rescue.

There are also the unexpected and unclassified calls, which no amount of training can anticipate. In the summer of 1999, for example, Flemming was ordered to meet an American nuclear submarine on secret patrol in the North Atlantic about twenty-two kilometres from the coast. As the *Sambro* drew near, Flemming's crew saw that the sub had just surfaced, and the first person to emerge from the top of the conning tower was a sailor with a machine gun at the ready. The sub's hull was soon swarming with crew members up for some fresh air, while several men with Geiger counters checked the exposed deck for stray radiation. Flemming ordered the *Sambro*'s Zodiac into the water, and it soon returned with a sailor suffering from a ruptured appendix. The man had been stripped of all his U.S. Navy insignia and had to decline Flemming's offer to swap logo-bearing gear such as ball caps and shirts. "Man, I'd love to, but I've got nothing on me that would identify me with that ship," he told Flemming. The patient, who had little confidence in the medical specialist on board the submarine, was delivered to the Canadian Navy dockyards in Halifax. He was discreetly treated, while the sub carried on with its undisclosed mission.

Then there was the Swissair disaster. The sturdy *Sambro* arrived with no clear idea among crew members about what to expect. They all grew sick at the overwhelming stench of jet fuel. And for five brutal days, the deck was continually loaded and unloaded with scrap metal, personal effects and human remains. Flemming was not aboard but took charge of a small fast-rescue craft, which was itself soon piled high with debris, remains and a dozen soggy passports lined up along the windshield. Concerned about his crew handling too much dangerous debris, Flemming successfully experimented with a cargo net draped across the open deck that could be used to hoist everything at once aboard bigger collector ships such as the *Preserver*. Flemming also brought in some trout-fishing nets to capture more of the small floating debris and body parts.

There are exciting technological breakthroughs on the horizon for search and rescue in Canada. Remote sensing by satellite will someday play a vital role in locating individual survivors, perhaps identifying the wisps of heat from their bodies. Experiments are under way with remote-controlled rescue craft that can be dropped by air and then manoeuvred to the site of a marine disaster. Cellphones will be equipped with Global Positioning System chips to automatically provide longitude and latitude whenever users call for help. Well-equipped aircraft such as the new Cormorant rescue helicopters will pierce the worst weather, from fog to severe icing conditions. Survival gear is becoming more sophisticated, effective and affordable. But technology will never provide more than tools that can break down, give false information or need jury-rigging. The heart of search and rescue will always be the people who return again and again to those terrifying, life-threatening frontiers. People like Ken Hill, Tony Rodgers, Paul Jackman, Tony Isaacs, Darcy St-Laurent, Tim Eagle, Chris Flemming and hundreds of others whose names never appear in newspapers or on television newscasts. Collectively they have helped make Canada a centre of excellence in the rarefied world of search and rescue, a place others visit to learn how it's done in the worst environments the planet has to offer. They have built long-standing reputations as the best in the business. More important to all of them, they have pulled hundreds of men, women and children back from the brink, restoring the gift of life itself.

Acknowledgments

Interviews with dozens of key players in the search-and-rescue business provided much of the information in *Deadly Frontiers*. Many of the conversations were for background purposes only and have not been credited in the text. I was also given permission to observe training — and the occasional mission — on Labradors, Hercules, Sea Kings, frigates and coast guard vessels, and I was given tours of the rescue centres in Trenton, Ontario, Halifax, Nova Scotia, and St. John's, Newfoundland, as well as the search-and-rescue training facilities of the Canadian Coast Guard college near Sydney, Nova Scotia. The Department of National Defence allowed me to visit the National Search and Rescue Secretariat office and library in Ottawa and the defence research labs in Hull, Quebec. I was given the run of 413 Transport and Rescue Squadron at Greenwood, Nova Scotia, on several occasions and sat in on numerous briefings and meetings. Many documents cited in the text were obtained through the federal Access to Information Act and Nova Scotia's freedom of information legislation, while others came from public sources or through informal requests.

The published literature on search and rescue in Canada is thin and uneven. The best written and researched book on an aircraft crash is Robert Mason Lee's *Death and Deliverance: The Haunting True Story of the Hercules Crash at the North Pole* (1992), which deals at some length with that problematic rescue operation. Peter Tadman's *The Survivor* (1991) tells the story of the 1972 Marten Hartwell crash. *Heartbreak and Heroism: Canadian Search and Rescue Stories* (1997), by John Melady, gathers more than a dozen post-war stories about search and rescue. *Mayday: The Perils of the Waves* (1998), by Nicholas Faith, based on a British television series, has sections on the *Flare* disaster and the sinking of the *Gold Bond Conveyor*. Michael Hirsh ably tells the story of the *Salvador Allende* disaster in *Pararescue* (2001). In 1994, the Para Rescue Association of Canada published an ambitious collection of stories, facts and figures on search and rescue as seen through the eyes of search-and-rescue technicians, who were then celebrating the half-century mark of their profession; *That Others May Live: 50 Years of Para Rescue in Canada*, a mix of valuable research and trade gossip, was published in a limited edition and is quite difficult to find. The Swissair crash has already resulted in at least two books, Stephen Kimber's *Flight 111: The Tragedy of the Swissair Crash* (1999) and Don Ledger's *Swissair Down: A Pilot's View of the Crash at Peggy's Cove* (2000). Both authors understandably treat the search-and-rescue aspect of the disaster as peripheral to the main story.

Special thanks go to Colonel Randy Price and Lieutenant-Colonel Mike Dorey, who kindly allowed me to watch the operations of 413 Squadron from the inside and were themselves a rich source of wisdom about the evolution of search and rescue; to Master Corporal Darcy St-Laurent and Sergeant Tim Eagle, Sartechs who endured my constant questioning with candor and good humour; to Major Jim Armour, who helped me understand the complexities of the Labrador 305 investigation; to Glenn Chamberlain, who reliably provided information on dozens of active searches at the Halifax rescue co-ordination centre; to Tony Rodgers, who taught me just what it is that groundpounders do; and to Ken Hill, who patiently guided me through the complexities of ground search theory. Many others helped in many ways, and not all of their names appear in the text. I hope this book becomes a token of my profound gratitude by bringing the remarkable story of their profession to a wider public.

I flew with several of the men who died in the 1998 Marsoui crash during the previous summer at Greenwood — some of those trips aboard Labrador 305 — as a visiting journalist trying to learn the complicated business of search and rescue. One bright July day, Captain Darrin Vanderbilche was co-pilot as our crew did a leisurely training mission in a Labrador over the Maritimes. We had done some hoisting routines from the centre hatch over a field along the Prince Edward Island shore. As we crossed Nova Scotia's Minas Basin on the way back to Greenwood, I remember Vanderbilche asking the aircraft commander whether we could divert slightly so he could have an aerial view of an Annapolis Valley house he was thinking of buying near the base. The radio chatter among the crew was upbeat, full of real-estate advice for someone with so full a future. Vanderbilche talked excitedly about maybe putting in a pool. It was a golden moment as a tightly meshed crew of young men relaxed after a day of hard training. I also especially remember Master Corporal Darrell Cronin, a lean and muscular Newfoundlander who was so tall his pants always seemed a little stretched. Cronin clearly had little time for nosy journalists and remained aloof for several days. But one day, while training aboard a bouncing Hercules, he took an interest in making sure I was well harnessed and comfortable, giving me a few decisive thumbs-up after checking all the straps and buckles. As the aircraft headed back to base on a glorious sunny afternoon, the crew kept the rear ramp of the Hercules open for an extra fifteen minutes so the cargo-hold passengers could enjoy some aerial sightseeing. Darrell and I sat on an emergency pallet fastened to the ramp and contentedly watched the glory of Nova Scotia roll by slowly a kilometre below. It was a moment of pure peace that Darrell and his crewmates now live with forever.

My colleagues at the Canadian Press news agency have been a continuing source of inspiration and have helped this project along in numerous ways, tangible and otherwise. Judy Monchuk was a faithful co-pilot during CP Halifax's coverage of most of the disasters discussed in this book, including *Flare*, Swissair and the crash of Tusker 27. Her advice, insight and good humour over the years have helped to keep my own journalism fresh and relevant. Stephen Thorne, now in Ottawa, shared many search-and-rescue ideas and interviews with me; indeed, we reported together on many a rescue story in some of the most exacting journalism of our careers. Other CP stalwarts helped in many other ways, including Michelle MacAfee and Michael MacDonald, who cheerfully turned over tapes and other material; my Toronto bosses, Eric Morrison and Scott White, who gave me the freedom to explore new ways of doing journalism, including use of freedom-of-information laws; Peter Buckley, whose high standards have haunted me long after his retirement; and everyone in the CP Halifax bureau, past and present, who has pursued excellence in this difficult profession.

Thanks must go as well to Laurel Boone at Goose Lane Editions, who was a keen, supportive and perceptive editor; and to Ryan Astle, Goose Lane's skilled graphic designer. I especially want to thank an older, retired gentleman who visited Halifax a few years back and showed me the navy building that once was home to the military's rescue co-ordination centre. Dennis had been a young navigator on the RCAF's Lancaster bombers in the early 1950s stationed at Summerside, Prince Edward Island, and Greenwood, Nova Scotia, but discovered to his dismay that he suffered from chronic air sickness. Each flight was an hours-long nightmare of nausea in cramped quarters. He urgently requested a humanitarian transfer to a ground job and was accommodated with a posting to the overnight shift at the Halifax rescue centre, then located in a brick building near the railway station in the city's south end. Many of the overnight crew — my retiree included — often slept all the night through, knowing that a junior staffer would rouse them if something serious happened. At the time, the RCAF had a limited search-and-rescue role, responding to far fewer incidents with meagre resources. The napping overnight crew was always well rested and able to pursue with vigour many personal daytime activities, such as courting the young ladies of Halifax. Dennis did just that. He married one of the Conway girls, and they soon had a blue-eyed baby boy. That baby was me, and so began my tenuous attachment to the world of search and rescue. You could say that, were it not for search and rescue, I might not be here today to tell these stories.

Dean Beeby
Halifax, Nova Scotia

Index

C

S

MURRAY BREWSTER

Dean Beeby was born in Halifax and grew up in Edmonton, Kingston, Kitchener, Dorval, and Toronto. With a master's degree in history from the University of Toronto, he became a journalist. In 1990, he returned to Halifax, where he is chief of the Canadian Press's Atlantic service.

An ardent advocate and user of freedom-of-information laws, Beeby has spoken at numerous conferences of journalists and international policy-makers. He co-edited *Moscow Dispatches: Inside Cold War Russia* and is the author of *In a Crystal Land: Canadian Explorers in Antarctica* and *Cargo of Lies: The True Story of a Nazi Double Agent in Canada*.